CRC Exam Mastery 2025-2026

600+ Practice Questions for Certified Risk Adjustment Coder Success with HCC Scenarios, RADV Audit Prep & Detailed Explanations

Edwin Eldon Wolak

ISBN: 978-1-7642339-6-5

Isohan Publishing

Table of Contents

Chapter 1: The Risk Adjustment Revolution

The healthcare payment system stands at a crossroads. Medicare Advantage plans now cover more than half of all Medicare beneficiaries, and this shift represents far more than changing enrollment numbers—it signals a fundamental transformation in how we deliver and pay for healthcare. Risk adjustment coding sits at the heart of this revolution, determining billions of dollars in payments and shaping the future of patient care.

You might wonder why risk adjustment matters so much. The answer lies in the numbers: Medicare Advantage enrollment has surged from 54% to an projected 64% by 2034, creating a $590 billion market that depends entirely on accurate coding. Every diagnosis you capture, every condition you document, and every code you assign directly impacts patient care funding and health plan sustainability.

Medicare Advantage Growth Creates Unprecedented Opportunities

The Medicare Advantage expansion represents the largest shift in healthcare delivery since the creation of Medicare itself. Plans that once served niche markets now compete for millions of beneficiaries, each bringing unique health conditions and care needs. This growth creates both challenges and opportunities for skilled risk adjustment professionals.

Current enrollment data shows **27.4 million beneficiaries** chose Medicare Advantage in 2024, up from just 5.3 million in 2005. The Centers for Medicare & Medicaid Services projects this number will reach **35 million by 2034**, representing nearly two-thirds of all Medicare-eligible Americans. Each new enrollee brings an average of **3.2 chronic conditions**, creating exponential demand for accurate risk adjustment coding.

Health plans recognize that survival depends on precise risk capture. Plans with superior coding accuracy maintain competitive advantage through better risk score optimization, while those with poor coding practices face financial penalties and potential market exit. This reality drives unprecedented investment in risk adjustment infrastructure, creating career opportunities across the coding spectrum.

Consider the mathematics: a single missed diabetes complication can cost a health plan **$3,200 annually** in lost revenue. Multiply this across thousands of members, and the financial impact becomes staggering. Plans that achieve 95% coding accuracy versus 85% accuracy can realize millions in additional revenue—money that directly funds member benefits and care improvements.

The $590 Billion Payment Model Revolution

Risk adjustment transforms healthcare economics by linking payments to patient acuity rather than simple head counts. This paradigm shift requires deep understanding of how diagnoses translate into revenue, and why certain conditions generate higher payments than others.

The Hierarchical Condition Category (HCC) model creates payment differentials based on expected healthcare costs. A healthy 65-year-old generates approximately **$9,500** in annual payments, while a member with diabetes, heart failure, and chronic kidney disease can generate over **$25,000** annually. These payment variations reflect actual care costs and ensure plans receive appropriate funding for sicker populations.

Payment accuracy depends on precise diagnosis capture and documentation. The current CMS-HCC Model V28 recognizes **115 payment categories**, each with specific coding requirements and hierarchical relationships. Understanding these relationships becomes essential for maximizing appropriate reimbursement while maintaining compliance standards.

Case Example 1: The Diabetic Complexity Challenge

Maria Gonzalez, a 72-year-old Medicare Advantage member, visits her primary care physician for routine diabetes management. Her medical record shows:

- Type 2 diabetes mellitus with diabetic nephropathy
- Chronic kidney disease, stage 3
- Hypertension
- Diabetic retinopathy with macular edema

The novice coder might capture only the basic diabetes code (E11.9), missing significant revenue opportunities. The skilled risk adjustment coder recognizes multiple HCC opportunities:

- HCC 35 (Diabetes with acute complications) for the nephropathy
- HCC 326 (Chronic kidney disease, stage 3-4) for the CKD
- HCC 238 (Hypertensive heart and kidney disease) for complicated hypertension
- Additional HCC 122 (Proliferative diabetic retinopathy) for the eye complications

This comprehensive coding increases Maria's risk score from 1.2 to 2.8, generating an additional **$15,200** in annual payments. These funds support enhanced diabetes management programs, specialty care coordination, and preventive services that improve Maria's long-term outcomes.

RADV Audit Transformation Demands Excellence

The Risk Adjustment Data Validation (RADV) audit program represents the most significant compliance challenge facing Medicare Advantage plans. Beginning in 2025, CMS will audit **550 plans annually**, examining medical records to verify the accuracy of submitted diagnosis codes. This represents a 400% increase from current audit levels and fundamentally changes the risk adjustment compliance environment.

RADV audits focus on high-dollar diagnosis categories and frequently challenged conditions. Plans with poor documentation practices face substantial financial penalties, with some organizations reporting **70% failure rates** on audited diagnoses. These failures result in payment recoupment, penalty assessments, and potential enrollment sanctions.

The audit process examines three critical elements: medical record availability, provider qualifications, and documentation adequacy. Each diagnosis must meet specific MEAT criteria (Monitor, Evaluate, Assess, Treat) and demonstrate active management during the payment year. Historical documentation practices that satisfied traditional coding requirements often fail RADV standards.

Successful organizations invest heavily in audit preparation, including provider education, documentation improvement, and coding quality programs. These investments create career opportunities for professionals who understand both clinical documentation requirements and regulatory compliance standards.

Case Example 2: The RADV Documentation Failure

Westside Medical Group received a RADV audit notification for member James Patterson, whose submitted diagnoses included:

- Chronic systolic heart failure (HCC 221)
- Chronic kidney disease, stage 4 (HCC 326)
- Major depressive disorder (HCC 155)

The audit revealed critical documentation deficiencies:

- Heart failure mentioned only in problem list without clinical assessment
- CKD stage based on single lab value without physician interpretation
- Depression noted in medication list but not addressed in encounter notes

These documentation failures resulted in **$18,500** in payment recoupment for a single member. The plan's response required extensive provider retraining, new documentation protocols, and enhanced coding review processes. This case demonstrates why organizations prioritize candidates with RADV audit experience and documentation expertise.

Career Pathways in the New Risk Adjustment Economy

The risk adjustment career landscape offers unprecedented opportunities for motivated professionals. Entry-level positions typically start at **$45,000-$55,000** annually, while experienced risk adjustment directors earn **$120,000-$180,000** in major markets. Specialized roles in audit preparation, compliance, and analytics command premium salaries.

Career progression follows predictable patterns: entry-level coder, senior risk adjustment specialist, coding supervisor, manager, and director. Each level requires additional skills and knowledge, but the investment pays substantial returns. Many professionals advance from entry-level to management positions within 3-5 years through focused skill development and certification achievement.

Remote work opportunities abound in risk adjustment, with many organizations offering fully distributed teams. This flexibility attracts career changers and provides work-life balance often missing in traditional healthcare roles. Geographic location becomes less important than skill level and performance outcomes.

Consulting opportunities emerge for experienced professionals who understand both clinical and technical aspects of risk adjustment. Independent consultants often earn **$100-$200** per hour, working with plans, providers, and technology vendors on implementation projects and compliance initiatives.

Industry Challenges Create Professional Opportunities

The risk adjustment industry faces several persistent challenges that create opportunities for skilled professionals. Provider education remains inadequate, with many physicians lacking basic understanding of documentation requirements. Health plans struggle to balance aggressive coding practices with compliance requirements. Technology solutions often fail to address workflow realities and clinical complexities.

Documentation quality varies dramatically across provider types and specialties. Primary care physicians, who generate the majority of risk adjustment diagnoses, often receive minimal training in HCC concepts. Specialists focus on clinical outcomes rather than coding implications. These educational gaps create opportunities for professionals who can bridge clinical and coding perspectives.

Health plans invest millions in technology solutions that promise automated coding and documentation improvement. However, these systems require skilled professionals who understand both technical capabilities and clinical workflows. Implementation success depends on professionals who can translate business requirements into operational realities.

Case Example 3: The Multi-System Documentation Challenge

Regional Health Plan implemented a new clinical documentation improvement (CDI) program across 200 primary care providers. The goal: improve risk score capture while maintaining compliance standards.

Initial results showed concerning patterns:

- 35% of diabetes patients lacked complication documentation
- Heart failure severity remained unspecified in 60% of cases
- Mental health conditions appeared in medication lists but not encounter diagnoses

The CDI team, led by experienced risk adjustment professionals, developed targeted intervention strategies:

- Provider-specific feedback reports highlighting missed opportunities
- Clinical decision support tools integrated into electronic health records
- Educational webinars focusing on high-value diagnosis categories
- Audit simulation exercises to demonstrate documentation requirements

Six months post-implementation, the plan achieved a **12% improvement** in risk score capture while maintaining 98% RADV audit pass rates. This success required professionals who understood clinical workflows, coding requirements, and change management principles.

The Foundation for Success

Risk adjustment coding demands more than traditional medical coding skills. Success requires understanding payment methodologies, clinical documentation principles, audit requirements, and compliance standards. The most successful professionals develop expertise across multiple domains while maintaining focus on accuracy and integrity.

Educational investment pays substantial returns in this field. The Certified Risk Adjustment Coder (CRC) credential serves as the industry standard, but additional certifications in compliance, auditing, and healthcare analytics enhance career prospects. Continuous learning becomes essential as regulations change and industry practices advance.

Networking within the risk adjustment community provides access to job opportunities, best practice sharing, and professional development resources. Industry conferences, online forums, and professional associations offer platforms for skill development and relationship building.

The risk adjustment revolution creates both challenges and opportunities for healthcare professionals. Those who master the technical skills, understand the business implications, and maintain ethical standards will find themselves at the center of healthcare's most significant transformation. The question isn't whether you should enter this field—it's how quickly you can develop the skills necessary to succeed.

Setting the Stage for Mastery

This book provides the knowledge, tools, and strategies necessary for risk adjustment success. Each chapter builds upon previous concepts while introducing new complexities. Practice questions reinforce learning while case studies demonstrate real-world applications. The goal: transform you from a curious newcomer into a confident risk adjustment professional.

The Medicare Advantage revolution continues to accelerate, creating urgent demand for skilled professionals who can navigate its complexities. Health plans that master risk adjustment will thrive, while those that struggle will face financial pressures and potential market exit. Your role in this transformation depends on your commitment to learning, growing, and adapting to an ever-changing environment.

Risk adjustment represents more than a coding specialty—it embodies the future of healthcare payment and delivery. Mastering this field positions you at the forefront of healthcare transformation, contributing to better patient outcomes while building a rewarding career. The revolution has begun, and your journey starts now.

Key Principles to Remember:

- Medicare Advantage growth creates unprecedented career opportunities in risk adjustment
- Payment accuracy depends on precise diagnosis capture and thorough documentation
- RADV audits demand new levels of compliance and documentation excellence
- Career advancement requires continuous learning and skill development across multiple domains
- Success combines technical expertise with business understanding and ethical practice

Chapter 2: Risk Adjustment Models Decoded

Healthcare payment models operate on complex mathematical foundations that determine how billions of dollars flow through the system. Risk adjustment models translate patient diagnoses into payment predictions, creating the financial framework that supports modern healthcare delivery. Understanding these models requires more than memorizing formulas—it demands comprehension of the clinical, actuarial, and policy decisions that shape healthcare economics.

The transition from simple capitation to sophisticated risk adjustment represents one of healthcare's most significant innovations. Rather than paying identical amounts for all patients, risk adjustment recognizes that healthcare costs vary dramatically based on patient characteristics, chronic conditions, and care needs. This recognition fundamentally changed how healthcare organizations approach patient care, resource allocation, and financial planning.

CMS-HCC Model V28 Architecture and Payment Logic

The Centers for Medicare & Medicaid Services Hierarchical Condition Category Model Version 28 represents the culmination of decades of actuarial analysis and policy refinement. This model processes millions of Medicare claims annually to predict future healthcare costs based on patient demographics and diagnosed conditions. The current model recognizes **115 distinct payment categories**, each representing a cluster of related diagnoses with similar cost implications.

Model V28 incorporates several design principles that distinguish it from previous versions. The hierarchy concept prevents double-counting of related conditions, ensuring that patients receive payment for their most severe manifestation of each disease process. Interaction terms capture the additional costs associated with multiple chronic conditions. Demographic factors account for age and gender-related cost variations.

The model operates through a sophisticated coefficient system that assigns relative weights to each HCC category. These weights, derived from analysis of actual Medicare expenditures, reflect the additional costs associated with each condition compared to a baseline healthy population. For example, acute myocardial infarction (HCC 160) carries a coefficient of 0.323, meaning it adds approximately $3,100 to annual payment expectations for the average Medicare beneficiary.

Constrained coefficients represent a critical model feature that prevents inappropriate payment inflation. Certain condition combinations receive lower combined weights than the sum of individual coefficients, reflecting clinical realities where multiple conditions may share common treatment pathways or represent different manifestations of the same underlying disease process.

Understanding RAF Score Calculations

The Risk Adjustment Factor (RAF) score provides the mathematical foundation for all Medicare Advantage payments. This score represents the relative costliness of an individual beneficiary compared to the average Medicare population, which receives a RAF score of 1.0. Scores above 1.0 indicate higher expected costs, while scores below 1.0 suggest lower cost expectations.

RAF score calculation follows a structured process:

1. Identify all eligible diagnoses from the previous year's medical encounters
2. Map diagnoses to their corresponding HCC categories
3. Apply hierarchical relationships to eliminate duplicate conditions
4. Sum the coefficients for remaining categories
5. Add demographic factors based on age, gender, and eligibility status
6. Apply any interaction terms for multiple chronic conditions
7. Adjust for institutional vs. community dwelling status

Case Example 1: Complex RAF Score Calculation

Margaret Thompson, age 78, presents with multiple chronic conditions documented during the previous calendar year. Her diagnoses include:

- Type 2 diabetes with diabetic nephropathy (maps to HCC 35)
- Chronic kidney disease, stage 4 (maps to HCC 326)
- Congestive heart failure, systolic (maps to HCC 221)
- Major depressive disorder (maps to HCC 155)
- Chronic obstructive pulmonary disease (maps to HCC 276)

The RAF calculation process:

- Demographic factor (female, age 78): 0.453
- HCC 35 (Diabetes with complications): 0.318

- HCC 326 (Chronic kidney disease stage 4-5): 0.237
- HCC 221 (Congestive heart failure): 0.323
- HCC 155 (Major depressive disorder): 0.309
- HCC 276 (COPD): 0.328
- Interaction term (diabetes/CHF): 0.154

Margaret's total RAF score: 0.453 + 0.318 + 0.237 + 0.323 + 0.309 + 0.328 + 0.154 = **2.122**

This score indicates Margaret's expected healthcare costs are approximately 112% higher than the average Medicare beneficiary. At an average county rate of $12,000 annually, Margaret generates approximately **$25,464** in capitation payments to her Medicare Advantage plan.

HHS-HCC Model for Marketplace Populations

The Department of Health and Human Services HCC model serves commercial marketplace populations under the Affordable Care Act. While sharing conceptual similarities with the CMS-HCC model, HHS-HCC addresses different population characteristics, benefit structures, and regulatory requirements. Understanding these differences becomes essential for organizations serving both Medicare and commercial populations.

The HHS-HCC model recognizes **127 condition categories**, reflecting the broader age range and different morbidity patterns found in commercial populations. Unlike Medicare's focus on chronic conditions in older adults, the marketplace model must account for acute conditions, pregnancy-related costs, and pediatric conditions that rarely appear in Medicare data.

Risk adjustment transfers in the marketplace operate differently than Medicare Advantage payments. Rather than receiving prospective capitation adjustments, marketplace plans participate in a retrospective transfer system that redistributes funds from plans with lower-risk populations to those serving higher-risk members. This transfer mechanism creates different financial incentives and operational requirements.

The marketplace model incorporates cost-sharing reduction (CSR) factors that account for enhanced benefits provided to lower-income enrollees. These adjustments recognize that identical medical conditions generate different total costs depending on member cost-sharing obligations and utilization patterns.

CDPS and Specialized Risk Adjustment Models

The Chronic Illness and Disability Payment System (CDPS) serves Medicaid populations in many states, addressing the unique characteristics of dual-eligible and disabled populations. CDPS recognizes that traditional commercial risk factors may inadequately predict costs in populations with severe mental illness, developmental disabilities, and complex social determinants of health.

CDPS employs a hierarchical structure similar to HCC models but emphasizes different condition categories. Mental health conditions receive enhanced weighting, reflecting their significant impact on Medicaid costs. Developmental disabilities and complex medical conditions that rarely appear in commercial populations receive appropriate recognition.

Several states have developed customized risk adjustment models that incorporate social determinants of health, geographic factors, and state-specific population characteristics. These models recognize that standardized approaches may inadequately address local population needs and healthcare delivery patterns.

Private payer risk adjustment models vary significantly across organizations but generally follow HCC principles while incorporating proprietary adjustments. Some models emphasize pharmaceutical costs, others focus on utilization patterns, and many incorporate social determinants data unavailable in traditional HCC models.

Case Example 2: Model Comparison Across Populations

David Martinez, age 45, carries identical diagnoses across three different insurance programs:

- Major depressive disorder, severe
- Type 2 diabetes with diabetic nephropathy
- Hypertension

Medicare Advantage (CMS-HCC Model V28):

- Age/gender demographic factor: 0.379
- HCC 155 (Major depressive disorder): 0.309
- HCC 35 (Diabetes with complications): 0.318
- Total RAF: 1.006

ACA Marketplace (HHS-HCC Model):

- Age/gender factor: 0.645
- HCC 058 (Major depressive disorder): 0.414
- HCC 017 (Diabetes with complications): 0.318
- Total risk score: 1.377

Medicaid (CDPS Model):

- Age/gender factor: 0.523
- Mental health category weight: 0.892
- Diabetes category weight: 0.445
- Total risk score: 1.860

The variation across models reflects different population baselines, benefit structures, and policy objectives. David's condition generates the highest risk score under Medicaid CDPS, reflecting that program's emphasis on mental health conditions and the relative rarity of his condition combination in the Medicaid population baseline.

Model Transitions and Implementation Challenges

Risk adjustment models undergo periodic updates to reflect changing medical practice, new treatments, and evolving healthcare delivery patterns. Model transitions create significant operational challenges for health plans, requiring system updates, staff retraining, and financial impact analysis.

The transition from CMS-HCC Model V24 to V28 exemplifies these challenges. The new model eliminated several payment categories, modified others, and introduced new hierarchical relationships. Plans experienced varying financial impacts depending on their member populations, with some seeing payment increases while others faced reductions.

Blended payment years ease the transition impact by gradually phasing in new model coefficients over multiple years. During these transition periods, payments reflect weighted averages of old and new model results, providing time for health plans to adapt their operations and for the model to demonstrate stability.

Model calibration requires continuous refinement based on actual cost experience and emerging medical evidence. CMS regularly publishes model updates, coefficient adjustments, and technical corrections that require immediate implementation. Organizations must maintain robust change management processes to address these ongoing modifications.

Advanced Model Analytics and Performance Optimization

Successful risk adjustment requires sophisticated analytical capabilities that extend beyond basic coding accuracy. Organizations must understand member-level risk score trends, identify coding gaps, and predict financial impacts of various scenarios. These capabilities require investment in technology, personnel, and analytical processes.

Predictive modeling helps organizations identify members likely to develop new conditions or experience disease progression. These models combine traditional risk factors with social determinants data, utilization patterns, and clinical indicators to target intervention programs and coding initiatives.

Benchmark analysis compares organizational performance against industry standards and competitor organizations. These comparisons identify opportunities for improvement while highlighting areas of relative strength. Understanding competitive positioning becomes essential for strategic planning and resource allocation.

Case Example 3: Performance Analytics in Action

Mountain View Health Plan analyzed their risk adjustment performance across multiple dimensions:

Member-Level Analysis:

- Average RAF score: 1.24 (industry benchmark: 1.31)
- RAF score trend: +3.2% annually (benchmark: +4.1%)
- Documentation gaps: 15% of members missing chronic condition updates

Provider-Level Analysis:

- Primary care RAF capture: 89% of potential (benchmark: 93%)
- Specialist RAF capture: 76% of potential (benchmark: 82%)
- Documentation quality scores: 78% meeting MEAT criteria

Financial Impact Analysis:

- Estimated revenue opportunity: $4.2 million annually
- Documentation improvement ROI: 7:1 over two years
- Coding accuracy investment: $800,000 annually

These analytics guided targeted interventions including provider education programs, enhanced clinical decision support, and expanded coding review processes. The plan achieved a 8.5% improvement in RAF capture within 18 months while maintaining compliance standards.

Technology Integration and Automation Opportunities

Risk adjustment operations increasingly rely on sophisticated technology solutions that automate routine tasks while highlighting exceptions requiring human intervention. Natural language processing systems scan clinical documentation to identify potential HCC diagnoses, while machine learning algorithms predict member risk score trends and identify coding gaps.

Electronic health record integration provides real-time coding feedback to providers, suggesting diagnosis codes and documentation improvements during patient encounters. These systems reduce administrative burden while improving coding accuracy and compliance.

Artificial intelligence applications include automated medical record review, coding suggestion engines, and fraud detection systems. While these technologies show promise, they require careful implementation and ongoing oversight to ensure accuracy and compliance with regulatory requirements.

The most successful organizations combine technology capabilities with human expertise, using automation to enhance rather than replace professional judgment. Technology handles routine tasks while professionals focus on complex cases, quality assurance, and strategic initiatives.

Building Model Expertise for Career Success

Risk adjustment model expertise requires continuous learning and adaptation to changing requirements. Successful professionals develop deep understanding of model mathematics while maintaining practical knowledge of implementation challenges and operational realities.

Professional development should encompass actuarial concepts, statistical analysis, and healthcare economics in addition to traditional coding skills. Understanding how models translate into business outcomes enables more effective communication with leadership and enhanced career advancement opportunities.

Industry engagement through professional organizations, conferences, and peer networks provides access to emerging trends and best practices. The risk adjustment community actively shares knowledge and experiences, creating opportunities for learning and collaboration.

Certification programs provide structured learning paths and professional recognition. The CRC credential represents the industry standard, while additional certifications in analytics, compliance, and healthcare economics enhance professional value.

Mastering the Mathematical Foundation

Risk adjustment models represent sophisticated mathematical constructs that determine healthcare payment flows worth hundreds of billions of dollars annually. These models translate clinical diagnoses into financial predictions, creating the economic framework that supports modern healthcare delivery. Success in risk adjustment requires understanding both the technical mechanics and practical applications of these complex systems.

The evolution from simple capitation to sophisticated risk adjustment reflects healthcare's ongoing maturation and the recognition that one-size-fits-all payment approaches inadequately address population health needs. As these models continue to advance, professionals who master their complexities will find themselves at the center of healthcare's most significant transformations.

Model expertise extends beyond memorizing coefficients and calculating RAF scores. True mastery requires understanding the clinical logic underlying model construction, the policy objectives driving model design, and the operational challenges of model implementation. This comprehensive understanding enables professionals to contribute meaningfully to organizational success while advancing their own career objectives.

Essential Model Mastery Elements:

- CMS-HCC Model V28 recognizes 115 payment categories with specific hierarchical relationships
- RAF scores translate diagnoses into payment predictions through sophisticated coefficient systems
- Multiple risk adjustment models serve different populations with varying objectives and methodologies
- Model transitions require careful change management and ongoing analytical support

- Technology solutions enhance but do not replace the need for professional expertise and judgment

Chapter 3: ICD-10-CM Mastery for Risk Adjustment

Medical coding in the risk adjustment context demands precision that extends far beyond traditional reimbursement requirements. Each diagnosis code assignment can generate thousands of dollars in annual payments, making accuracy both financially critical and ethically essential. The ICD-10-CM classification system provides the foundation for all risk adjustment activities, but success requires understanding which codes generate payments, how specificity affects outcomes, and what documentation supports code assignment.

The transition from ICD-9-CM to ICD-10-CM fundamentally changed risk adjustment coding by providing unprecedented diagnostic specificity. Where ICD-9-CM offered approximately 14,000 diagnosis codes, ICD-10-CM provides over 70,000 codes with enhanced anatomical detail, severity indicators, and clinical specificity. This expansion creates both opportunities and challenges for risk adjustment professionals who must navigate increased complexity while maintaining accuracy standards.

The Universe of Risk-Adjusting Diagnoses

Of the more than 70,000 available ICD-10-CM codes, approximately **7,770 diagnoses map to HCC payment categories**. This subset represents conditions with significant impact on healthcare costs and resource utilization. Understanding which codes generate payments becomes essential for effective risk adjustment coding, but equally important is recognizing that non-mapping codes often provide clinical context necessary for appropriate code selection.

HCC-mapping codes fall into distinct clinical categories that reflect common chronic conditions, acute complications, and complex medical situations. Cardiovascular conditions represent the largest category, followed by diabetes complications, mental health disorders, and respiratory conditions. Each category requires specific clinical knowledge and documentation understanding.

The Centers for Medicare & Medicaid Services publishes annual updates to the code-to-HCC crosswalk that reflect new codes, revised definitions, and clinical advances. These updates require immediate attention from risk adjustment professionals who must ensure their coding practices remain current and compliant.

Hierarchy relationships between HCC categories affect code selection strategies. More severe manifestations of disease processes supersede less severe categories, meaning that coders must understand clinical progressions and documentation requirements for various severity levels.

Case Example 1: Diabetes Coding Complexity

Eleanor Washington, a 68-year-old Medicare Advantage member, presents for diabetes management. Her medical record contains the following information:

- "Type 2 diabetes with diabetic nephropathy"
- "Chronic kidney disease, stage 3"
- "Diabetic retinopathy without macular edema"
- "Peripheral neuropathy"

The coding analysis reveals multiple pathway options:

Option 1 - Basic Coding:

- E11.9 (Type 2 diabetes mellitus without complications)
- N18.3 (Chronic kidney disease, stage 3)
- H36.9 (Retinal disorder, unspecified)
- G62.9 (Polyneuropathy, unspecified)

Option 2 - Risk Adjustment Optimized:

- E11.21 (Type 2 diabetes mellitus with diabetic nephropathy)
- N18.3 (Chronic kidney disease, stage 3)
- E11.319 (Type 2 diabetes mellitus with unspecified diabetic retinopathy without macular edema)
- E11.40 (Type 2 diabetes mellitus with diabetic neuropathy, unspecified)

The optimized coding approach maps to:

- HCC 35 (Diabetes with acute complications) - coefficient 0.318
- HCC 326 (Chronic kidney disease, stage 3-4) - coefficient 0.237
- Additional diabetes complications recognition

This coding improvement generates approximately **$5,340** in additional annual payments while accurately reflecting Eleanor's clinical complexity and care requirements.

Official Coding Guidelines Application in Risk Adjustment

The ICD-10-CM Official Guidelines for Coding and Reporting provide the regulatory foundation for all diagnosis code assignment. These guidelines take precedence over any organizational policies or external guidance, making thorough understanding essential for compliant risk adjustment coding.

Several guideline sections prove particularly relevant for risk adjustment applications. Section I.C.4 addresses diabetes coding requirements, including the presumptive relationship between diabetes and certain complications. Section I.C.9 covers circulatory system conditions, including heart failure specificity requirements. Section I.C.10 addresses respiratory conditions, particularly chronic obstructive pulmonary disease classification.

The "code also" and "use additional code" instructional notes require careful attention in risk adjustment coding. These notes often identify opportunities for additional HCC capture while ensuring complete clinical picture representation. Ignoring these instructions can result in incomplete coding and missed revenue opportunities.

Combination codes present both opportunities and challenges in risk adjustment coding. These codes capture multiple related conditions in a single code assignment, often providing more comprehensive documentation than separate code assignments. However, understanding when combination codes are appropriate versus when separate codes are required becomes essential for optimal results.

Specificity Requirements and Clinical Documentation

Risk adjustment coding demands maximum specificity supported by clinical documentation. Unspecified codes rarely map to HCC categories, and when they do, they typically generate lower payments than their specified counterparts. This reality creates strong incentives for enhanced clinical documentation and specific diagnosis capture.

Laterality requirements for certain conditions affect both code assignment and payment outcomes. While many conditions do not require laterality specification for HCC mapping, others demand precise anatomical identification. Understanding these requirements prevents coding errors and ensures appropriate payment capture.

Severity indicators built into ICD-10-CM codes often determine HCC mapping eligibility. Heart failure codes exemplify this principle, with systolic, diastolic, and combined heart failure each generating different payment outcomes. Coders must understand these distinctions and ensure documentation supports the assigned severity level.

Episode of care indicators affect certain condition categories, particularly injuries and poisonings. Initial encounters, subsequent encounters, and sequela encounters may generate different payment outcomes or no payments at all. Understanding these distinctions prevents inappropriate code assignment and potential compliance issues.

Case Example 2: Specificity Impact on Payment Outcomes

Robert Chen, age 74, receives treatment for chronic obstructive pulmonary disease. The provider documentation states: "COPD exacerbation with acute respiratory failure."

Less Specific Coding:

- J44.1 (Chronic obstructive pulmonary disease with acute exacerbation)
- J96.00 (Acute respiratory failure, unspecified with hypoxia or hypercapnia)

This coding maps to HCC 276 (COPD) with coefficient 0.328.

More Specific Coding:

- J44.1 (Chronic obstructive pulmonary disease with acute exacerbation)
- J96.00 (Acute respiratory failure, unspecified with hypoxia or hypercapnia)

Additional documentation review reveals:

- Arterial blood gas showing CO2 retention
- Bipap initiation for respiratory support

Revised Specific Coding:

- J44.1 (Chronic obstructive pulmonary disease with acute exacerbation)
- J96.02 (Acute respiratory failure with hypercapnia)

The addition of specificity regarding hypercapnia maps to HCC 207 (Respirator dependence/tracheostomy status) in addition to HCC 276, generating an additional **$2,970** annually while accurately reflecting Robert's clinical complexity.

Common Coding Errors and Avoidance Strategies

Risk adjustment coding errors fall into predictable patterns that organizations can address through targeted education and quality assurance programs. Understanding these common mistakes enables professionals to develop prevention strategies and quality improvement initiatives.

Undercoding represents the most frequent error type, where coders select less specific codes than documentation supports. This pattern often reflects traditional coding habits focused on claim payment rather than risk adjustment optimization. Undercoding directly impacts revenue while potentially misrepresenting patient acuity.

Overcoding, while less common, creates significant compliance risks and potential audit exposure. Overcoding occurs when code assignments exceed documentation support or reflect conditions not present during the encounter. These errors can result in payment recoupment, penalties, and potential fraud allegations.

Hierarchy violations occur when coders assign multiple codes within the same HCC hierarchy, failing to recognize that only the most severe category receives payment. These errors indicate insufficient understanding of HCC relationships and can complicate audit responses.

Historical versus current condition confusion represents another common error pattern. Risk adjustment requires evidence of active management during the payment year, but coders sometimes assign codes for resolved conditions or historical diagnoses without current relevance.

Specialty-Specific Coding Considerations

Different medical specialties present unique coding challenges and opportunities in risk adjustment. Primary care providers generate the majority of HCC diagnoses but often lack specialized knowledge of complex coding requirements. Specialists possess detailed clinical knowledge but may not understand risk adjustment implications of their documentation choices.

Cardiology encounters frequently involve complex conditions with multiple severity levels and complication possibilities. Understanding the distinctions between systolic and diastolic heart failure, acute versus chronic manifestations, and complication versus co-existing condition becomes essential for appropriate coding.

Endocrinology visits often center on diabetes management with multiple potential complications. Coders must understand the relationships between diabetes, kidney disease, eye complications, and neuropathy while recognizing when conditions represent complications versus separate disease processes.

Oncology coding requires understanding of cancer staging, treatment status, and remission classification. Active treatment generates different HCC categories than history of cancer, and documentation must clearly indicate current status for appropriate code assignment.

Psychiatry and behavioral health coding demands understanding of condition severity, episode classification, and comorbidity relationships. Mental health conditions often interact with substance abuse disorders, creating complex hierarchical relationships.

Case Example 3: Multi-Specialty Coding Coordination

Grace Martinez, age 71, receives care from multiple specialists for complex medical conditions:

Primary Care Visit:

- "Diabetes type 2, well controlled"
- "Hypertension, stable"
- "Depression, improved on medication"

Endocrinology Visit:

- "Type 2 diabetes mellitus with diabetic nephropathy"
- "Chronic kidney disease, stage 3, stable"
- "Diabetic retinopathy, recommend ophthalmology follow-up"

Cardiology Visit:

- "Chronic systolic heart failure, EF 35%"
- "Coronary artery disease with previous MI"
- "Hypertensive heart disease"

Nephrology Visit:

- "Chronic kidney disease, stage 3, secondary to diabetes"

- "Metabolic acidosis secondary to CKD"
- "Anemia of chronic kidney disease"

Coordinated coding captures the complete clinical picture:

- E11.21 (Type 2 DM with diabetic nephropathy) - HCC 35
- N18.3 (Chronic kidney disease, stage 3) - HCC 326
- I50.20 (Unspecified systolic heart failure) - HCC 221
- I25.2 (Old myocardial infarction) - HCC 160
- E11.319 (Type 2 DM with diabetic retinopathy) - additional support for HCC 35
- N18.6 (End stage renal disease) - if progression occurs

This comprehensive approach generates a RAF score of approximately 2.4, representing **$28,800** in annual payments that support Grace's complex care coordination across multiple specialties.

Technology Tools and Coding Support Systems

Modern risk adjustment coding benefits from sophisticated technology tools that enhance accuracy while improving efficiency. Computer-assisted coding systems scan clinical documentation to identify potential HCC diagnoses, providing coding suggestions based on documented clinical findings.

Natural language processing engines analyze unstructured clinical notes to extract relevant diagnostic information and suggest appropriate code assignments. These systems require careful implementation and ongoing oversight to ensure accuracy and compliance with coding guidelines.

Electronic health record integration provides real-time coding feedback during clinical documentation, alerting providers to potential HCC opportunities and documentation deficiencies. These systems improve both coding accuracy and clinical documentation quality.

Coding quality monitoring systems track accuracy rates, identify error patterns, and provide targeted feedback to individual coders. These systems enable continuous improvement while ensuring consistency across coding teams.

Advanced Coding Strategies for Risk Adjustment Excellence

Successful risk adjustment coding requires strategies that extend beyond basic guideline compliance. Advanced practitioners develop systematic approaches to documentation review, code selection, and quality assurance that maximize accuracy while minimizing compliance risks.

Documentation mining involves systematic review of complete medical records to identify all supportable HCC diagnoses. This approach requires understanding of clinical workflows, provider documentation patterns, and regulatory requirements.

Code validation processes ensure that assigned codes meet both ICD-10-CM guidelines and risk adjustment requirements. These processes include clinical review, documentation verification, and compliance assessment.

Provider education programs help clinical staff understand the connection between documentation quality and accurate risk representation. These programs focus on practical documentation improvements rather than coding-specific training.

Quality assurance metrics track coding accuracy across multiple dimensions, including guideline compliance, specificity levels, and HCC capture rates. These metrics enable continuous improvement while identifying training needs.

Building Excellence Through Systematic Approaches

ICD-10-CM mastery for risk adjustment requires more than memorizing code descriptions and guidelines. Success demands understanding of clinical concepts, payment methodologies, compliance requirements, and quality assurance principles. The intersection of these domains creates opportunities for professionals who can navigate complexity while maintaining accuracy and integrity.

The evolution of risk adjustment coding continues to accelerate, with new codes, revised guidelines, and enhanced technology tools creating both opportunities and challenges. Professionals who master the fundamental principles while adapting to changing requirements will find themselves well-positioned for career advancement and professional success.

Excellence in risk adjustment coding ultimately serves patients by ensuring appropriate funding for their care while supporting health plan sustainability. This dual responsibility requires professionals who can balance aggressive capture with compliance requirements, creating value for all stakeholders while maintaining the highest ethical standards.

Core Coding Excellence Principles:

- Master the 7,770 ICD-10-CM codes that map to HCC payment categories and their clinical contexts
- Apply official coding guidelines with precision while understanding risk adjustment implications
- Develop specialty-specific expertise that addresses unique documentation and coding challenges
- Implement systematic quality assurance processes that ensure accuracy and compliance
- Leverage technology tools to enhance rather than replace professional coding judgment

Chapter 4: Documentation Excellence

Clinical documentation serves as the foundation for all risk adjustment activities, bridging the gap between patient care and payment accuracy. Every diagnosis code, every payment calculation, and every audit response ultimately depends on the quality and completeness of medical record documentation. Understanding what constitutes excellent documentation requires knowledge of clinical requirements, regulatory standards, and practical implementation strategies.

The shift from traditional fee-for-service documentation to risk adjustment-focused documentation represents a fundamental change in healthcare record-keeping. While traditional documentation primarily supported individual encounter billing, risk adjustment documentation must paint a complete picture of patient complexity that supports ongoing payment adjustments and withstands regulatory scrutiny.

MEAT Criteria Fundamentals

The MEAT criteria (Monitor, Evaluate, Assess, Treatment) establish the minimum documentation standards for risk adjustment coding. These criteria ensure that documented diagnoses reflect active clinical management rather than passive acknowledgment of historical conditions. Understanding and applying MEAT criteria correctly can mean the difference between successful audits and costly payment recoupment.

Monitor requires evidence that providers actively track the condition through ongoing clinical observation. This monitoring can include vital sign trends, laboratory results, symptom assessment, functional status evaluation, or patient-reported outcomes. The key element is documented evidence that the provider maintains awareness of the condition's current status.

Evaluate demands evidence of provider assessment regarding the condition's impact on the patient's health status. This evaluation might include interpretation of diagnostic tests, assessment of treatment effectiveness, comparison to previous status, or consideration of the condition's interaction with other health issues.

Assess requires clinical decision-making regarding the condition's significance and management approach. This assessment includes treatment planning, prognosis consideration, risk stratification, or care coordination decisions. The provider must demonstrate active thought processes regarding the condition's management.

Treatment encompasses all interventions related to the condition, including pharmaceutical management, therapeutic interventions, patient education, care coordination, or monitoring schedules. Treatment evidence can be direct intervention or deliberate decisions to continue, modify, or discontinue existing therapies.

Case Example 1: MEAT Criteria Application

Martha Rodriguez, age 76, presents for routine primary care follow-up. Her chart contains multiple chronic conditions requiring MEAT documentation:

Chronic Systolic Heart Failure:

- **Monitor:** "Patient reports stable functional capacity, no recent shortness of breath. Last echo showed EF 35% unchanged from prior."
- **Evaluate:** "Current symptoms suggest stable compensated heart failure with good medication adherence."
- **Assess:** "Will continue current ACE inhibitor and beta-blocker regimen. Consider cardiology referral if symptoms worsen."
- **Treatment:** "Continue lisinopril 10mg daily and metoprolol 50mg twice daily. Follow-up in 3 months."

Type 2 Diabetes with Nephropathy:

- **Monitor:** "HbA1c today 7.2%, improved from 8.1% six months ago. Creatinine stable at 1.8."
- **Evaluate:** "Diabetes control improving with current regimen. Nephropathy appears stable without progression."
- **Assess:** "Current medications effective. Will continue monitoring kidney function given diabetic nephropathy."
- **Treatment:** "Continue metformin 1000mg twice daily. Nephrology follow-up scheduled for next month."

This documentation clearly supports both conditions for risk adjustment purposes while demonstrating comprehensive clinical management.

Face-to-Face Encounter Requirements

Risk adjustment regulations require that HCC diagnoses originate from face-to-face encounters between patients and qualified healthcare providers. These requirements ensure that diagnoses reflect current clinical assessment rather than administrative coding exercises. Understanding encounter requirements becomes essential for compliant risk adjustment practices.

Acceptable face-to-face encounters include office visits, hospital encounters, emergency department visits, urgent care visits, and certain home health encounters. Telemedicine visits qualify when they include real-time audio and video communication, though documentation requirements remain identical to in-person encounters.

The timing of encounters affects risk adjustment eligibility. Diagnoses must be documented during encounters occurring within the payment year (January 1 through December 31) to qualify for that year's risk adjustment. This timing requirement creates urgency around annual wellness visits and chronic disease management appointments.

Encounter frequency does not affect individual diagnosis eligibility, meaning that a single qualifying encounter can support a diagnosis for the entire payment year. However, organizations often encourage multiple encounters per year to strengthen audit responses and demonstrate ongoing clinical management.

Documentation timing relative to the encounter affects compliance status. While providers have some flexibility in completing documentation after encounters, the documentation must reflect clinical findings and decisions made during the actual encounter. Retrospective additions or modifications may create compliance risks.

Acceptable Provider Types and Credentials

Risk adjustment regulations specify which healthcare providers can generate qualifying diagnoses for HCC purposes. These provider type requirements ensure that diagnoses reflect appropriate clinical expertise while preventing inappropriate gaming of the risk adjustment system.

Physicians represent the primary source of risk adjustment diagnoses, including medical doctors (MDs), doctors of osteopathic medicine (DOs), and specialists in all recognized medical specialties. These providers can document and support diagnoses within their scope of practice and clinical expertise.

Non-physician practitioners (NPPs) including nurse practitioners, physician assistants, and clinical nurse specialists can generate risk adjustment diagnoses when operating within their scope of practice and under appropriate supervision where required. Their documentation must meet identical standards to physician documentation.

Certain healthcare providers cannot generate risk adjustment diagnoses despite their clinical qualifications. These excluded providers typically include physical therapists, occupational therapists, speech therapists, and other allied health professionals whose scope of practice focuses on specific interventions rather than comprehensive medical assessment.

Provider supervision requirements vary by state and specialty but generally require that non-physician providers operate under appropriate physician oversight when treating Medicare beneficiaries. These requirements affect both encounter validity and documentation requirements.

Case Example 2: Provider Type Documentation Scenarios

Regional Medical Group employs various provider types who encounter Medicare Advantage members:

Qualifying Encounter - Family Medicine Physician: Dr. Sarah Chen documents: "68-year-old male with Type 2 diabetes mellitus with diabetic nephropathy, chronic kidney disease stage 3. Reviewing today's labs showing creatinine 1.9, stable from prior. Continue ACE inhibitor, monitor kidney function quarterly."

Qualifying Encounter - Nurse Practitioner: Sarah Johnson, NP documents: "Follow-up for chronic systolic heart failure. Patient reports stable symptoms, no recent hospitalizations Physical exam shows clear lungs, no peripheral edema. Echo last month showed EF 40%. Continue current heart failure medications."

Non-Qualifying Encounter - Physical Therapist: Mike Thompson, PT documents: "Patient with diabetes and heart failure tolerating exercise program well. No chest pain or shortness of breath during therapy session. Will continue current exercise prescription."

The physician and nurse practitioner documentation generates qualifying risk adjustment diagnoses, while the physical therapist documentation, despite being clinically relevant, cannot support HCC diagnoses due to provider type restrictions.

Clinical Documentation Improvement Integration

Clinical Documentation Improvement (CDI) programs play an increasingly important role in risk adjustment success. These programs bridge the gap between clinical care and documentation requirements, helping providers create records that accurately reflect patient complexity while meeting regulatory standards.

CDI professionals review medical records to identify documentation gaps, unclear clinical indicators, and missed diagnosis opportunities. Their work focuses on accuracy and completeness rather than coding optimization, ensuring that documentation improvements reflect genuine clinical complexity.

Provider queries represent a key CDI tool for clarifying ambiguous documentation. These queries ask providers to specify unclear clinical findings, confirm suspected diagnoses, or provide additional clinical context. Effective queries focus on clinical clarification rather than coding suggestions.

Real-time CDI interventions during patient encounters provide immediate feedback to providers regarding documentation opportunities. These interventions can include clinical decision support alerts, documentation templates, or concurrent chart review by CDI specialists.

Technology-enabled CDI programs use natural language processing and machine learning to identify documentation patterns and predict improvement opportunities. These systems require careful implementation to ensure they enhance rather than replace professional clinical judgment.

Documentation Quality Metrics and Monitoring

Successful documentation improvement requires systematic measurement and monitoring of quality indicators. These metrics help organizations identify improvement opportunities, track progress, and ensure sustainable documentation practices.

Specificity rates measure the percentage of diagnoses documented with maximum available detail. Higher specificity rates typically correlate with improved risk adjustment outcomes and better audit performance. Tracking specificity by provider and condition type identifies targeted improvement opportunities.

MEAT criteria compliance rates indicate the percentage of risk adjustment diagnoses supported by appropriate clinical documentation. These rates should approach 100% for conditions included in risk adjustment submissions, as deficient documentation creates audit vulnerabilities.

Provider documentation consistency measures variation in documentation practices across similar clinical scenarios. High variation suggests training opportunities and the need for standardized documentation approaches.

Audit simulation results provide early warning of potential compliance issues by applying RADV audit criteria to current documentation. These simulations help organizations identify and address weaknesses before facing actual audits.

Case Example 3: Documentation Quality Improvement Initiative

Metro Health Plan implemented a comprehensive documentation quality program across their provider network:

Baseline Assessment:

- Specificity rate: 72% (industry benchmark: 85%)
- MEAT criteria compliance: 68% (target: 95%)
- Provider variation: High (coefficient of variation: 0.34)

Intervention Strategies:

1. Provider education workshops focusing on high-value conditions
2. Real-time clinical decision support in electronic health records
3. Monthly provider-specific feedback reports
4. CDI specialist consultation for complex cases

Implementation Results (12 months):

- Specificity rate improved to 89%
- MEAT criteria compliance increased to 92%
- Provider variation decreased (coefficient: 0.18)
- Estimated additional revenue: $3.2 million annually

The program succeeded by combining education, technology, and feedback mechanisms while maintaining focus on clinical accuracy rather than revenue optimization alone.

Advanced Documentation Strategies

Excellence in risk adjustment documentation requires strategies that extend beyond minimum compliance requirements. Advanced practitioners develop systematic approaches that maximize accuracy while minimizing provider burden and maintaining clinical workflow efficiency.

Template development creates standardized documentation frameworks that ensure consistent capture of required elements while allowing for clinical customization. Effective templates include prompts for MEAT criteria elements without restricting provider flexibility or clinical judgment.

Provider education programs focus on the clinical rationale behind documentation requirements rather than coding-specific training. These programs help providers understand how documentation quality affects patient care funding and health plan sustainability.

Audit preparation processes include regular internal review of documentation using RADV criteria, proactive identification of documentation weaknesses, and systematic improvement of deficient areas. These processes reduce audit risk while improving overall documentation quality.

Technology integration includes clinical decision support tools that provide real-time documentation feedback, natural language processing systems that identify potential documentation gaps, and automated quality monitoring systems that track improvement metrics.

Overcoming Common Documentation Challenges

Healthcare organizations face predictable documentation challenges that can be addressed through systematic approaches and targeted interventions. Understanding these challenges enables proactive solution development and improved documentation outcomes.

Provider resistance to documentation changes often reflects concerns about workflow disruption, administrative burden, or lack of understanding regarding documentation importance. Successful change management addresses these concerns through education, workflow optimization, and clear communication about benefits.

Time constraints in clinical encounters create pressure to minimize documentation effort while maintaining quality. Solutions include documentation templates, voice recognition software, and streamlined data entry processes that reduce administrative burden.

Electronic health record limitations can impede effective documentation through poor interface design, inadequate prompt systems, or insufficient integration with clinical workflows. Organizations must work with vendors to optimize systems for risk adjustment documentation requirements.

Inconsistent provider training creates variation in documentation quality and compliance levels. Standardized training programs, regular reinforcement sessions, and competency assessments help ensure consistent documentation practices across provider networks.

Creating a Culture of Documentation Excellence

Outstanding risk adjustment documentation results from organizational cultures that value accuracy, completeness, and continuous improvement. These cultures recognize that documentation quality directly affects patient care funding while supporting provider satisfaction and organizational sustainability.

Leadership commitment to documentation excellence creates the foundation for successful improvement initiatives. This commitment includes resource allocation, priority setting, and consistent messaging about documentation importance across all organizational levels.

Provider engagement in documentation improvement ensures that initiatives address real workflow challenges while maintaining clinical relevance. Successful programs involve providers in solution development rather than imposing external requirements without clinical input.

Continuous learning approaches recognize that documentation requirements evolve continuously due to regulatory changes, clinical advances, and organizational growth. These approaches include ongoing education, regular metric review, and adaptive improvement strategies.

Quality assurance processes provide ongoing monitoring and feedback regarding documentation performance. These processes include regular auditing, provider-specific feedback, and systematic correction of identified deficiencies.

The foundation of risk adjustment success rests on clinical documentation that accurately captures patient complexity while meeting regulatory standards. This documentation serves multiple purposes: supporting appropriate payment, enabling quality improvement, and facilitating care coordination. Professionals who master documentation requirements while helping providers create excellent records will find themselves essential to organizational success.

Documentation Excellence Framework:

- MEAT criteria provide the foundation for all risk adjustment documentation and must be consistently applied
- Face-to-face encounters with qualified providers generate the only acceptable documentation for HCC purposes
- Clinical Documentation Improvement programs bridge the gap between clinical care and documentation requirements
- Quality metrics and monitoring systems enable continuous improvement and proactive compliance management
- Advanced documentation strategies maximize accuracy while minimizing provider burden and maintaining workflow efficiency

Chapter 5: HCC Categories Deep Dive - Cardiovascular & Metabolic

Cardiovascular and metabolic conditions represent the cornerstone of risk adjustment coding, accounting for the highest payment categories and affecting the largest number of Medicare Advantage members. These conditions create complex coding scenarios where clinical nuance directly translates into payment outcomes, making precise understanding essential for both accuracy and compliance. The interplay between heart disease, diabetes, kidney disease, and their various complications creates a web of coding relationships that demands both clinical knowledge and technical expertise.

The cardiovascular system generates more HCC categories than any other body system, reflecting both the prevalence of heart disease in Medicare populations and the significant cost impact of these conditions. Understanding the distinctions between acute and chronic manifestations, systolic versus diastolic dysfunction, and compensated versus decompensated states becomes critical for appropriate code assignment and optimal risk adjustment outcomes.

Heart Disease HCC Categories 221-227

The heart disease HCC categories represent some of the highest-value diagnosis groups in risk adjustment, with coefficients ranging from 0.323 for acute myocardial infarction to 0.368 for cardio-respiratory failure. These categories require precise clinical understanding and meticulous documentation to ensure appropriate capture and audit defense.

HCC 221 - Acute Myocardial Infarction captures both ST-elevation and non-ST-elevation myocardial infarctions occurring within the current year. This category requires evidence of acute presentation and cannot be used for historical infarctions or chronic coronary conditions. Documentation must include specific timing, clinical presentation, and treatment interventions to support this high-value category.

HCC 222 - Unstable Angina and Other Acute Ischemic Heart Disease includes unstable angina, demand ischemia, and acute coronary syndromes without myocardial infarction. These conditions require evidence of acute presentation and active management, distinguishing them from stable coronary disease.

HCC 223 - Chronic Atherosclerosis encompasses stable coronary artery disease, previous myocardial infarction, and chronic ischemic conditions. This category provides ongoing payment for patients with established coronary disease requiring continued management and monitoring.

HCC 224 - Other Acute and Chronic Ischemic Heart Disease captures conditions not fitting other categories, including coronary dissection, thrombosis of coronary prostheses, and other specified ischemic conditions.

HCC 225 - Congestive Heart Failure represents one of the most frequently captured and highest-impact categories in risk adjustment. This category requires evidence of heart failure diagnosis with current symptoms or ongoing management.

HCC 226 - Cardio-Respiratory Failure and Shock captures the most severe cardiovascular conditions, including cardiogenic shock, acute cardio-respiratory failure, and severe decompensated heart failure requiring intensive interventions.

HCC 227 - Heart Failure with Reduced Ejection Fraction specifically addresses systolic heart failure with documented ejection fraction reduction. This category reflects the higher costs associated with reduced ejection fraction heart failure compared to preserved ejection fraction conditions.

Case Example 1: Complex Coronary Disease Management

William Thompson, age 72, presents to cardiology for ongoing coronary disease management. His complex cardiac history includes:

Current Documentation: "Patient returns for follow-up of ischemic cardiomyopathy with systolic heart failure, EF 30%. Previous anterior STEMI in 2019 with LAD stenting. Current symptoms include mild dyspnea on exertion, well-controlled on optimal medical therapy. Recent echo shows persistent LV dysfunction but no acute changes. Continue ACE inhibitor, beta-blocker, and aldosterone antagonist. Patient stable for discharge with cardiology follow-up in 3 months."

Coding Analysis:

- I25.5 (Ischemic cardiomyopathy) maps to HCC 223 (coefficient 0.273)
- I50.20 (Unspecified systolic heart failure) maps to HCC 225 (coefficient 0.323)
- Z95.1 (Presence of aortocoronary bypass graft) provides clinical context

Documentation Enhancement Opportunities: The provider could enhance documentation by specifying: "Chronic systolic heart failure, currently compensated" and "History of ST-elevation myocardial infarction with subsequent ischemic cardiomyopathy." This approach provides clearer clinical picture while supporting appropriate HCC capture.

William's conditions generate combined coefficients of 0.596, adding approximately **$7,152** to his annual risk score. The precise documentation enables successful audit defense while accurately reflecting his clinical complexity and ongoing care requirements.

Diabetes Complications and Constrained Coefficients

Diabetes represents one of the most complex areas in risk adjustment coding due to the intricate relationships between various complications and the application of constrained coefficients that prevent inappropriate payment inflation. Understanding these relationships requires both clinical knowledge of diabetic complications and technical understanding of HCC hierarchy rules.

HCC 35 - Diabetes with Acute Complications captures diabetes with nephropathy, neurological complications, circulatory complications, and other specified complications. This category requires evidence of active diabetic complications rather than diabetes alone.

HCC 36 - Diabetes with Chronic Complications includes diabetes with ophthalmic manifestations, oral complications, and other chronic manifestations. The distinction between acute and chronic complications affects payment outcomes and audit requirements.

HCC 37 - Diabetes with Multiple Complications applies when patients have both acute and chronic diabetic complications documented. This category receives higher payment recognition for increased complexity.

HCC 38 - Diabetes without Complications provides payment for diabetes without documented complications but still requires evidence of active management and monitoring

The constrained coefficient system prevents double-counting when diabetes complications overlap with other condition categories. For example, diabetic nephropathy combined with chronic kidney disease receives a reduced combined coefficient rather than the sum of individual category payments.

Chronic Kidney Disease Staging and Documentation

Chronic kidney disease staging creates specific HCC categories based on severity levels, with higher stages generating increased payments reflecting greater care complexity and cost requirements. Understanding staging criteria and documentation requirements becomes essential for appropriate capture.

HCC 326 - Chronic Kidney Disease, Stage 3-4 requires documentation of moderate to severe kidney function reduction with specific staging information. Providers must document the stage based on laboratory values, clinical assessment, or nephrology consultation.

HCC 327 - Chronic Kidney Disease, Stage 5 captures end-stage renal disease requiring dialysis or transplantation consideration. This category generates significant payment increases reflecting intensive care requirements.

HCC 328 - End-Stage Renal Disease applies to patients receiving dialysis or having received kidney transplantation. This category requires evidence of current treatment status rather than historical conditions.

Staging documentation must include specific stage identification rather than general chronic kidney disease mentions. Laboratory values alone do not establish staging without provider interpretation and clinical correlation.

Case Example 2: Diabetic Complications with Kidney Disease

Maria Santos, age 68, presents for endocrinology follow-up with complex diabetic complications:

Clinical Documentation: "Type 2 diabetes mellitus with diabetic nephropathy, currently managed with ACE inhibitor and dietary modifications. Recent labs show creatinine 2.1, eGFR 28, indicating chronic kidney disease stage 4. Diabetic retinopathy noted on recent ophthalmology exam with recommendation for continued monitoring. Patient also reports numbness in feet consistent with diabetic peripheral neuropathy. Current HbA1c 7.8%, improved from previous visit."

Comprehensive Coding:

- E11.21 (Type 2 diabetes with diabetic nephropathy) → HCC 35
- N18.4 (Chronic kidney disease, stage 4) → HCC 326
- E11.319 (Type 2 diabetes with unspecified diabetic retinopathy) → HCC 35 support

- E11.40 (Type 2 diabetes with diabetic neuropathy) → HCC 35 support

Constrained Coefficient Application: Individual coefficients: HCC 35 (0.318) + HCC 326 (0.237) = 0.555 Constrained coefficient for diabetes/CKD combination: 0.486 Net payment impact: **$5,832** annually

This case demonstrates how constrained coefficients prevent payment inflation while ensuring appropriate recognition of legitimate clinical complexity. The documentation supports multiple diabetic complications while recognizing the interconnected nature of diabetic kidney disease.

Advanced Cardiovascular Coding Scenarios

Complex cardiovascular patients often present multiple overlapping conditions that require sophisticated understanding of HCC relationships, hierarchy rules, and clinical documentation requirements. These scenarios test both clinical knowledge and technical coding expertise.

Acute-on-chronic presentations create particular coding challenges where patients with chronic conditions experience acute exacerbations. These situations may qualify for multiple HCC categories or require selection of the most appropriate acute category based on clinical presentation and treatment focus.

Post-procedural complications and device-related conditions add complexity to cardiovascular coding. Understanding when complications generate separate HCC categories versus when they represent expected outcomes of underlying conditions becomes essential for accurate coding and compliance.

Comorbidity interactions between cardiovascular and other conditions create additional payment opportunities through interaction terms in the risk adjustment model. These interactions recognize that multiple chronic conditions create greater than additive cost impacts.

Case Example 3: Multi-System Cardiovascular Complexity

Robert Kim, age 74, presents to the emergency department with acute decompensation:

Emergency Department Documentation: "74-year-old male with known ischemic cardiomyopathy, EF 25%, presents with acute systolic heart failure exacerbation. Patient reports 3-day history of increasing shortness of breath, orthopnea, and bilateral lower extremity edema. Physical exam reveals elevated JVP, bilateral rales, and 3+ pitting edema. Chest X-ray shows pulmonary congestion. BNP elevated at 2,400. Initiated IV furosemide with good response."

Hospital Course Documentation: "Patient responded well to diuretic therapy with resolution of pulmonary congestion and improvement in symptoms. Echo shows unchanged EF 25% with global hypokinesis. Underlying coronary disease stable without acute changes. Discharged on optimized heart failure regimen including ACE inhibitor, beta-blocker, and aldosterone antagonist."

Coding Analysis:

- I50.21 (Acute systolic heart failure) → HCC 225 (0.323)
- I25.5 (Ischemic cardiomyopathy) → HCC 223 (0.273)
- Combined coefficient consideration for related conditions

Documentation Quality Assessment: The documentation clearly establishes acute exacerbation of chronic systolic heart failure with appropriate clinical findings, treatment response, and discharge planning. This supports both HCC 225 for the acute episode and ongoing chronic management requirements.

Metabolic Syndrome and Endocrine Complications

Metabolic conditions beyond diabetes create additional risk adjustment opportunities that require understanding of complex endocrine relationships and their impact on healthcare costs. These conditions often interact with cardiovascular disease to create multiplicative rather than additive cost impacts.

Obesity-related conditions generate payment adjustments when documented as morbid obesity or obesity with specific complications. Simple obesity documentation typically does not map to HCC categories, requiring evidence of complications or extreme severity.

Thyroid disorders create payment opportunities when documented with specific severity indicators or complications. Hypothyroidism and hyperthyroidism require evidence of active management and monitoring rather than stable, well-controlled conditions.

Adrenal disorders, including Addison's disease and Cushing's syndrome, generate significant payment adjustments due to their complexity and ongoing management requirements. These conditions require specialty care and careful monitoring, justifying enhanced payment recognition.

Technology Integration and Cardiovascular Coding

Advanced cardiovascular coding benefits from technology tools that can identify documentation patterns, suggest coding opportunities, and provide real-time feedback regarding clinical documentation quality. These tools must balance automation with clinical judgment to ensure accuracy and compliance.

Clinical decision support systems can alert providers to potential HCC opportunities based on documented clinical findings, laboratory values, and treatment patterns. These systems require careful calibration to avoid alert fatigue while maintaining clinical relevance.

Natural language processing can extract cardiovascular diagnostic information from unstructured clinical notes, identifying potential coding opportunities and documentation gaps. These systems require ongoing refinement to address medical terminology variations and clinical context.

Predictive analytics can identify patients likely to develop cardiovascular complications based on risk factors, utilization patterns, and clinical indicators. These tools enable proactive intervention and documentation improvement initiatives.

Quality Assurance and Audit Preparation

Cardiovascular conditions face intense scrutiny during RADV audits due to their high payment impact and frequent occurrence. Organizations must develop robust quality assurance processes that ensure documentation adequately supports assigned codes while maintaining clinical accuracy.

Pre-submission review processes should include clinical validation of cardiovascular diagnoses, verification of documentation adequacy, and assessment of hierarchy relationships. These reviews prevent submission errors while building confidence in audit outcomes.

Provider education specific to cardiovascular documentation should address common deficiencies, clarify clinical criteria, and provide practical examples of acceptable documentation. This education reduces coding errors while improving clinical documentation quality.

Mock audit exercises using cardiovascular cases help organizations identify documentation weaknesses and develop improvement strategies before facing actual audits. These exercises should use actual RADV criteria and scoring methodologies.

Internal audit programs should regularly review cardiovascular coding accuracy, identify trending issues, and implement corrective actions. These programs provide early warning of potential problems while supporting continuous improvement initiatives.

Mastering Cardiovascular Complexity

Cardiovascular and metabolic conditions represent the foundation of successful risk adjustment programs due to their prevalence, payment impact, and clinical complexity. Mastering these conditions requires integration of clinical knowledge, coding expertise, documentation skills, and quality assurance capabilities.

The relationships between various cardiovascular conditions, their interactions with metabolic disorders, and the application of constrained coefficients create a complex web that demands both technical precision and clinical understanding. Success in this area directly translates to organizational financial performance and audit compliance.

As cardiovascular care continues to advance and new treatment modalities emerge, risk adjustment professionals must maintain current knowledge while adapting their practices to reflect changing clinical realities. The professionals who master these complexities will find themselves essential to organizational success and well-positioned for career advancement.

Cardiovascular & Metabolic Mastery Elements:

- Heart disease HCC categories 221-227 represent the highest-value risk adjustment opportunities requiring precise clinical documentation
- Diabetes complications involve complex hierarchy relationships and constrained coefficients that prevent inappropriate payment inflation
- Chronic kidney disease staging creates specific payment categories based on severity levels with strict documentation requirements
- Cardiovascular coding scenarios often involve multiple overlapping conditions requiring sophisticated understanding of HCC relationships
- Quality assurance and audit preparation for cardiovascular conditions demand robust processes due to high payment impact and regulatory scrutiny

Chapter 6: HCC Categories - Neurological & Mental Health

Neurological and mental health conditions present unique challenges in risk adjustment coding due to their complex clinical presentations, evolving diagnostic criteria, and significant documentation requirements. These conditions often involve subtle clinical findings that require careful interpretation and precise code assignment to ensure appropriate payment while maintaining compliance standards. The intersection of neurological and psychiatric conditions creates additional complexity through comorbidity interactions and overlapping symptom presentations.

The expansion of neurological HCC categories in recent model updates reflects growing recognition of the cost impact associated with brain-related conditions and their long-term care requirements. Mental health conditions, while sometimes carrying lower individual coefficients, often interact with other medical conditions to create multiplicative cost impacts that justify enhanced payment recognition.

Expanded Neurological HCC Categories 193-196

The neurological HCC categories capture conditions affecting the central and peripheral nervous systems that require ongoing medical management and create significant healthcare utilization. These categories reflect both acute neurological events and chronic progressive conditions that demand specialized care and long-term monitoring.

HCC 193 - Hemiplegia/Hemiparesis captures paralysis or weakness affecting one side of the body, typically resulting from stroke, brain injury, or other neurological conditions. This category requires documentation of the specific anatomical distribution and functional impact of the neurological deficit.

HCC 194 - Monoplegia, Other Paralytic Syndromes includes isolated limb paralysis, quadriplegia, paraplegia, and other paralytic conditions not captured in other categories. These conditions require evidence of persistent neurological deficit with functional impact.

HCC 195 - Seizure Disorders and Convulsions encompasses epilepsy, recurrent seizures, and seizure disorders requiring ongoing anticonvulsant therapy. This category requires evidence of active seizure activity or ongoing treatment rather than remote seizure history.

HCC 196 - Coma, Brain Compression/Anoxic Damage captures the most severe neurological conditions including coma, anoxic brain injury, and conditions causing brain compression. These conditions typically require intensive care and have profound functional impacts.

These categories require careful documentation of clinical findings, functional impacts, and ongoing treatment requirements. Neurological conditions often involve progressive changes that require regular reassessment and documentation updates to maintain accurate risk adjustment.

Case Example 1: Stroke Sequelae Documentation

Helen Washington, age 71, presents for neurology follow-up six months after experiencing a left middle cerebral artery stroke:

Neurological Assessment: "Patient returns for stroke follow-up. Examination reveals persistent right-sided hemiparesis with 3/5 strength in right upper extremity and 4/5 strength in right lower extremity. Speech is clear without dysarthria or aphasia. Patient ambulates with walker, requires assistance with activities of daily living. MRI shows established left MCA infarct with no new changes. Continue physical therapy and occupational therapy. Adjust antiplatelet therapy and optimize stroke prevention measures."

Functional Assessment: "Patient demonstrates significant functional limitations related to right hemiparesis. Requires assistance with dressing, bathing, and transfers. Lives with daughter who provides daily care assistance. Physical therapy reports slow but steady improvement in mobility and strength."

Coding Considerations:

- G81.91 (Hemiplegia, unspecified affecting right dominant side) → HCC 193
- I69.351 (Hemiplegia following cerebral infarction affecting right dominant side) → HCC 193
- Z87.891 (Personal history of nicotine dependence) → Clinical context

The documentation clearly establishes persistent neurological deficits with functional impact, supporting HCC 193 with coefficient 0.436, generating approximately **$5,232** in additional annual payments. This payment supports ongoing rehabilitation services, specialized equipment, and enhanced care coordination required for stroke survivors.

Mental Health Condition Classification and Documentation

Mental health conditions in risk adjustment require precise diagnostic classification and evidence of active treatment to qualify for HCC recognition. The diagnostic criteria for psychiatric conditions continue to evolve, and documentation must reflect current clinical standards while supporting the assigned severity level.

Major Depressive Disorders require evidence of significant functional impairment and active treatment to qualify for HCC recognition. Documentation must distinguish between adjustment reactions, grief responses, and major depressive episodes requiring ongoing psychiatric intervention.

Bipolar Disorders encompass both manic and depressive episodes with evidence of mood cycling and functional impairment. Documentation should address current mood state, medication management, and psychosocial interventions.

Psychotic Disorders including schizophrenia and delusional disorders require evidence of thought disturbances, functional impairment, and ongoing psychiatric treatment. These conditions typically require specialty care and careful medication monitoring.

Anxiety Disorders qualify for HCC recognition when they create significant functional impairment requiring ongoing treatment. Simple anxiety or situational stress reactions typically do not meet HCC criteria without evidence of persistent symptoms and treatment requirements.

Mental health documentation must address both symptom presentation and functional impact to support HCC assignment. Treatment plans should include specific interventions, medication management, and monitoring requirements that demonstrate ongoing clinical attention.

Substance Abuse and Dependency Hierarchies

Substance abuse conditions create complex hierarchy relationships where more severe manifestations supersede less severe categories. Understanding these relationships prevents coding errors while ensuring appropriate payment for addiction treatment services.

Alcohol-Related Disorders range from abuse to dependence with various complication levels. Documentation must distinguish between current use disorders and historical alcohol problems to ensure appropriate code assignment.

Drug Dependence Disorders include opioid, cocaine, cannabis, and other substance dependencies with varying severity levels. These conditions require evidence of physiological dependence and treatment requirements rather than simple substance use.

Substance-Induced Disorders capture psychiatric conditions directly caused by substance use, including substance-induced psychosis, depression, and anxiety disorders. These conditions require clear causal relationships between substance use and psychiatric symptoms.

The hierarchy relationships prevent double-counting when patients have multiple substance-related conditions or when substance use disorders co-occur with substance-induced psychiatric conditions.

Case Example 2: Complex Mental Health Comorbidities

Michael Rodriguez, age 58, presents to psychiatry for ongoing mental health management:

Psychiatric Assessment: "Patient with long-standing bipolar I disorder, currently in mixed episode with prominent depressive features and mild manic symptoms. Reports medication compliance with lithium and quetiapine but continues to experience mood instability. Also managing alcohol use disorder, currently 8 months sober with AA attendance. Patient reports anxiety symptoms related to job stress and financial concerns."

Current Treatment Plan: "Adjust mood stabilizer dosing given persistent mood symptoms. Continue individual therapy and AA participation for alcohol use disorder. Consider intensive outpatient program if symptoms worsen. Patient demonstrates good insight and treatment engagement despite ongoing symptoms."

Functional Assessment: "Patient reports difficulty maintaining consistent work performance due to mood symptoms. Relationships strain during mood episodes. Has strong family support system and active engagement in recovery programs."

Coding Analysis:

- F31.31 (Bipolar disorder, current episode depressed, mild) → HCC 155
- F10.20 (Alcohol dependence, uncomplicated) → HCC 156
- Hierarchy consideration prevents double-counting of related conditions

This coding generates HCC 155 coefficient 0.309, adding approximately **$3,708** annually. The documentation supports active psychiatric treatment while demonstrating the ongoing complexity of managing multiple mental health conditions.

Neurological Manifestations and Progressive Conditions

Progressive neurological conditions create unique documentation challenges as they involve changing clinical presentations over time. These conditions require regular reassessment and updated documentation to maintain accurate risk adjustment while supporting appropriate care planning.

Dementia and Cognitive Disorders require evidence of cognitive impairment affecting daily functioning with documented clinical assessment. These conditions often progress over time, requiring regular cognitive assessments and care plan modifications.

Movement Disorders including Parkinson's disease, essential tremor, and dystonia require evidence of motor symptoms affecting daily functioning. Documentation should address both motor symptoms and any cognitive or psychiatric complications.

Multiple Sclerosis and other demyelinating diseases require evidence of neurological deficits with documented disease progression or relapse patterns. These conditions often involve multiple symptom domains requiring comprehensive assessment.

Peripheral Neuropathy conditions require evidence of nerve dysfunction with functional impact. Documentation should specify the underlying cause, anatomical distribution, and treatment requirements.

Case Example 3: Progressive Neurological Disease

Dorothy Chen, age 69, presents for neurology follow-up with advancing Parkinson's disease:

Neurological Examination: "Patient with Parkinson's disease diagnosed 5 years ago, now showing progression with increased rigidity and bradykinesia. Unified Parkinson's Disease Rating Scale score 42, increased from 28 last year. Tremor remains well-controlled with medications, but patient reports increased difficulty with fine motor tasks and balance issues."

Functional Assessment: "Patient reports increasing difficulty with buttoning clothes, writing, and walking without assistance. Recent fall last month without injury. Lives with spouse who provides increasing assistance with daily activities. Physical therapy evaluation recommended for gait training and fall prevention."

Treatment Plan: "Adjust carbidopa/levodopa dosing given disease progression. Add dopamine agonist for improved motor control. Refer to physical therapy and occupational therapy for functional assessment and intervention. Consider deep brain stimulation evaluation if medical management becomes inadequate."

Coding Considerations:

- G20 (Parkinson's disease) → HCC 79 (coefficient 0.454)
- Additional codes for complications as they develop

The documentation establishes progressive neurological disease with functional decline, supporting HCC 79 and generating approximately **$5,448** in additional annual payments to support enhanced care coordination and therapy services.

Psychiatric Documentation Standards and Treatment Evidence

Mental health conditions require specific documentation standards that demonstrate both clinical severity and ongoing treatment necessity. These standards ensure that HCC assignments reflect genuine psychiatric complexity rather than minor emotional disturbances or adjustment reactions.

Diagnostic criteria documentation should reference specific DSM-5 criteria when appropriate, demonstrating that the condition meets established psychiatric diagnostic standards. This documentation supports the clinical validity of assigned diagnoses during audit reviews.

Functional impairment assessment must demonstrate how psychiatric conditions affect daily living, work performance, social relationships, or other significant life areas. This assessment distinguishes between psychiatric conditions requiring HCC recognition and minor emotional problems.

Treatment plans should include specific interventions such as psychotherapy, medication management, case management, or intensive psychiatric services. These plans demonstrate ongoing treatment necessity and clinical attention required for complex psychiatric conditions.

Medication management documentation should address psychiatric medication selection, dosing adjustments, side effect monitoring, and treatment response assessment. This documentation demonstrates active psychiatric treatment and ongoing clinical complexity.

Interdisciplinary Coordination and Comprehensive Care

Neurological and mental health conditions often require coordinated care across multiple disciplines, creating documentation opportunities that demonstrate comprehensive treatment approaches. This coordination requires clear communication and shared care planning across providers.

Primary care integration with specialty psychiatric and neurological care creates opportunities for comprehensive documentation that captures the full scope of patient complexity. This integration requires clear communication regarding treatment plans and shared care responsibilities.

Allied health services including physical therapy, occupational therapy, speech therapy, and psychology provide additional documentation that supports the functional impact of neurological and psychiatric conditions. These services require coordination with medical management.

Social services and care management become essential for patients with severe neurological or psychiatric conditions affecting their ability to live independently. Documentation of these services demonstrates the comprehensive care requirements that justify enhanced payment.

Family involvement and caregiver support often become necessary for patients with progressive neurological conditions or severe mental illness. Documentation of family meetings, caregiver training, and support services demonstrates the broader care requirements.

Technology Tools for Neurological and Mental Health Coding

Specialized technology tools can enhance the accuracy and efficiency of neurological and mental health coding by identifying documentation patterns, suggesting clinical assessments and providing real-time feedback regarding coding opportunities.

Cognitive assessment tools integrated into electronic health records can prompt providers to document specific cognitive domains and functional impacts. These tools ensure comprehensive assessment while supporting appropriate HCC capture.

Mental health screening instruments can identify psychiatric symptoms requiring further evaluation and documentation. These instruments provide standardized approaches to psychiatric assessment while supporting clinical decision-making.

Neurological examination templates can ensure comprehensive documentation of neurological findings while prompting providers to address functional impacts and treatment plans. These templates improve both clinical care and documentation quality.

Clinical decision support systems can alert providers to potential drug interactions, medication monitoring requirements, and assessment needs specific to neurological and psychiatric conditions. These systems enhance patient safety while supporting comprehensive care documentation.

Audit Considerations for Neurological and Mental Health Conditions

Neurological and mental health conditions face specific scrutiny during RADV audits due to their subjective nature and documentation requirements. Organizations must prepare robust audit responses that demonstrate clinical validity and appropriate documentation support.

Objective clinical findings should be documented whenever possible for neurological conditions, including examination findings, diagnostic test results, and functional assessments. These objective findings strengthen audit responses when subjective symptoms are questioned.

Mental health conditions require clear documentation of diagnostic criteria, functional impairment, and treatment necessity. Audit responses should demonstrate that conditions meet established psychiatric diagnostic standards and require ongoing clinical attention.

Specialty consultation notes often provide the strongest audit support for complex neurological and psychiatric conditions. These consultations demonstrate appropriate clinical evaluation and support the validity of assigned diagnoses.

Treatment response documentation helps demonstrate the clinical necessity and ongoing nature of neurological and psychiatric conditions. Evidence of treatment modification, symptom monitoring, and functional assessment supports continued HCC recognition.

Navigating Neurological and Psychiatric Complexity

Neurological and mental health conditions require sophisticated understanding of clinical presentations, diagnostic criteria, and documentation requirements. These conditions often involve subtle findings that require careful clinical interpretation and precise documentation to ensure appropriate risk adjustment outcomes.

The expansion of neurological HCC categories reflects growing recognition of the cost and complexity associated with brain-related conditions. Mental health conditions, while sometimes carrying lower coefficients, create significant cost impacts through their interactions with other medical conditions and their effects on treatment compliance and health outcomes.

Success in neurological and mental health coding requires collaboration between clinical providers, coding professionals, and quality assurance teams to ensure that documentation accurately captures clinical complexity while meeting regulatory requirements. This collaboration becomes essential for both financial performance and audit compliance.

Neurological & Mental Health Excellence Framework:

- Expanded neurological HCC categories 193-196 capture conditions with significant functional impact requiring specialized documentation
- Mental health conditions require evidence of active treatment and functional impairment to qualify for HCC recognition
- Substance abuse hierarchies prevent double-counting while ensuring appropriate payment for addiction treatment complexity
- Progressive neurological conditions demand regular reassessment and updated documentation to maintain accuracy
- Interdisciplinary coordination creates comprehensive documentation that supports enhanced payment for complex care requirements

Chapter 7: HCC Categories Deep Dive - Respiratory, Immune & Other Systems

Respiratory conditions, immune disorders, and other complex medical conditions create significant risk adjustment opportunities while presenting unique documentation and coding challenges. These conditions often involve episodic exacerbations, complex medication regimens, and multisystem interactions that require sophisticated clinical understanding and precise documentation. The respiratory system alone accounts for multiple high-value HCC categories, while immune disorders and other conditions capture the clinical complexity of patients with rare diseases and complex medical needs.

The chronic nature of many respiratory conditions creates ongoing payment opportunities, but these conditions require evidence of active management and current clinical relevance rather than historical diagnoses. Immune disorders often involve expensive medications and specialist care that justify enhanced payment recognition, while other system conditions capture the remaining clinical complexity not addressed by major organ systems.

COPD Classification and Complexity Recognition

Chronic Obstructive Pulmonary Disease represents one of the most frequently captured HCC categories in Medicare Advantage populations, affecting approximately 15% of Medicare beneficiaries and generating substantial payment adjustments when properly documented. Understanding COPD classification and documentation requirements becomes essential for both revenue optimization and clinical accuracy.

HCC 276 - Chronic Obstructive Pulmonary Disease captures various forms of COPD including chronic bronchitis, emphysema, and mixed COPD presentations. This category requires evidence of airflow limitation with ongoing treatment requirements rather than simple smoking history or resolved respiratory conditions.

COPD documentation must include evidence of current disease activity through symptoms, medication requirements, pulmonary function testing, or acute exacerbations. Historical COPD diagnoses without current clinical relevance do not qualify for ongoing HCC recognition.

Exacerbation documentation creates additional coding opportunities when COPD patients experience acute worsening requiring enhanced treatment. These exacerbations often involve emergency department visits, hospitalizations, or significant medication adjustments that demonstrate disease complexity.

Severity indicators help differentiate between mild COPD requiring minimal intervention and severe COPD requiring continuous oxygen, frequent exacerbations, or advanced interventions. Documentation should reflect the current severity level and treatment requirements.

Case Example 1: COPD with Respiratory Failure

Frank Miller, age 74, presents to the emergency department with COPD exacerbation:

Emergency Department Assessment: "Patient with severe COPD presents with 3-day history of increased dyspnea, productive cough with purulent sputum, and decreased exercise tolerance. Physical examination reveals use of accessory muscles, prolonged expiratory phase, and diminished breath sounds with expiratory wheezes. Arterial blood gas shows pH 7.32, PaCO2 58, PaO2 62 on room air."

Treatment and Response: "Initiated bronchodilator therapy, systemic corticosteroids, and supplemental oxygen. Patient shows improvement with treatment but continues to require oxygen to maintain saturations above 90%. Admitted for continued monitoring and treatment of COPD exacerbation with acute respiratory failure."

Hospital Documentation: "Patient responded well to treatment with resolution of respiratory failure. Pulmonary function tests show severe airflow obstruction with FEV1 32% predicted. Discharged on oxygen therapy, increased bronchodilator regimen, and prednisone taper. Outpatient pulmonology follow-up arranged."

Coding Analysis:

- J44.1 (Chronic obstructive pulmonary disease with acute exacerbation) → HCC 276
- J96.02 (Acute respiratory failure with hypercapnia) → HCC 207
- Combined coefficients: 0.328 + 0.573 = 0.901

This comprehensive documentation generates **$10,812** in additional annual payments while accurately reflecting Frank's clinical complexity and ongoing care requirements. The documentation supports both the chronic COPD diagnosis and the acute respiratory failure complication.

Respiratory Failure Documentation and Acute Complications

Respiratory failure creates high-value HCC opportunities but requires precise documentation of clinical criteria and treatment requirements. Understanding the distinctions between various types of respiratory failure and their documentation requirements becomes essential for appropriate capture.

HCC 207 - Respirator Dependence/Tracheostomy Status captures patients requiring mechanical ventilation or having tracheostomy for respiratory support. This category requires evidence of ongoing dependency rather than temporary ventilation during surgical procedures.

Acute Respiratory Failure requires documentation of gas exchange abnormalities with specific criteria including oxygen requirements, arterial blood gas abnormalities, or clinical evidence of respiratory compromise. Simple shortness of breath or mild hypoxemia may not meet respiratory failure criteria.

Chronic Respiratory Failure involves ongoing gas exchange abnormalities requiring continuous treatment such as home oxygen therapy, non-invasive ventilation, or other respiratory support measures. Documentation must demonstrate chronic nature and ongoing treatment requirements.

Type I respiratory failure involves oxygenation problems without carbon dioxide retention, while Type II respiratory failure includes carbon dioxide retention. The distinction affects both clinical management and coding specificity requirements.

Autoimmune and Immune Disorder Recognition

Immune system disorders create complex coding scenarios due to their multisystem effects, expensive treatment requirements, and specialized care needs. These conditions often require biologic medications, frequent monitoring, and coordinated subspecialty care that justifies enhanced payment recognition.

Rheumatoid Arthritis and Inflammatory Arthritis conditions require evidence of active inflammation and ongoing treatment rather than historical diagnoses or joint symptoms alone. Documentation should include disease activity measures, treatment response, and functional impact assessment.

Systemic Lupus Erythematosus and other systemic autoimmune conditions affect multiple organ systems and require comprehensive documentation of disease manifestations, treatment requirements, and monitoring needs. These conditions often involve expensive medications and frequent laboratory monitoring.

Inflammatory Bowel Disease including Crohn's disease and ulcerative colitis requires evidence of active inflammation, treatment requirements, and disease complications. These conditions often involve surgery, biologic medications, and intensive monitoring.

Immunodeficiency Conditions both primary and secondary create increased infection risks and require ongoing monitoring and prophylactic treatments. Documentation should address the underlying cause, current immune status, and treatment requirements.

Case Example 2: Complex Autoimmune Disease

Susan Thompson, age 52, manages multiple autoimmune conditions requiring coordinated care:

Rheumatology Assessment: "Patient with established rheumatoid arthritis, currently active despite methotrexate and adalimumab therapy. Joint examination reveals synovitis in bilateral MCPs and PIPs with morning stiffness lasting 3 hours. DAS28 score 4.2 indicating moderate disease activity. Also managing secondary Sjögren's syndrome with sicca symptoms requiring artificial tears and saliva substitutes."

Current Treatment Plan: "Increase methotrexate to 20mg weekly with folic acid supplementation. Continue adalimumab injections. Added hydroxychloroquine for Sjögren's syndrome. Monitor CBC and liver function monthly due to immunosuppressive therapy. Ophthalmology follow-up for dry eye management."

Laboratory Monitoring: "Recent labs show mild leukopenia likely related to methotrexate therapy. RF positive, anti-CCP elevated. Complement levels normal. Monitoring for infection complications given immunosuppressive regimen."

Coding Considerations:

- M05.79 (Rheumatoid arthritis with rheumatoid factor, multiple sites) → HCC 58
- M35.00 (Sicca syndrome, unspecified) → Clinical context
- Z79.899 (Other long term drug therapy) → Biologic medication use

The documentation supports HCC 58 with coefficient 0.329, generating approximately **$3,948** annually while reflecting the complex care coordination and expensive treatments required for active autoimmune disease.

Cancer Diagnosis, Staging, and Treatment Status

Oncology coding in risk adjustment requires understanding of cancer staging, treatment phases, and the distinction between active cancer requiring treatment and history of cancer in remission. These distinctions significantly affect both payment outcomes and compliance requirements.

Active Cancer Diagnoses require evidence of current malignancy with ongoing treatment, monitoring, or progressive disease. These conditions generate substantial HCC payments reflecting the high costs of cancer care and treatment.

Cancer Staging affects HCC category assignment with metastatic cancers generating higher payments than localized cancers. Documentation must clearly indicate the extent of disease and staging information to support appropriate code assignment.

Treatment Status affects coding decisions with active treatment generating different coding outcomes than surveillance after completed treatment. Documentation should clarify current treatment plans and monitoring requirements.

History of Cancer may or may not qualify for HCC recognition depending on the cancer type, treatment completion status, and ongoing monitoring requirements. Recent cancer history often continues to generate HCC payments while remote history typically does not.

Transplant Status and Complications

Organ transplant recipients require ongoing immunosuppressive therapy and monitoring that justifies enhanced payment recognition through specific HCC categories. Understanding transplant-related coding requirements becomes essential for organizations serving transplant populations.

Transplant Status generates ongoing HCC recognition due to the lifelong immunosuppressive requirements and increased infection risks. Documentation must indicate current transplant status rather than historical transplant information alone.

Transplant Complications including rejection, graft-versus-host disease, and immunosuppression-related complications create additional coding opportunities. These complications require evidence of active treatment and ongoing management.

Immunosuppressive Therapy documentation should include specific medications, monitoring requirements, and complication management. These medications create ongoing costs and monitoring needs that justify enhanced payment.

Infection Risks in transplant recipients require ongoing monitoring and prophylactic treatments that demonstrate the ongoing complexity of transplant care. Documentation should address infection prevention strategies and monitoring protocols.

Case Example 3: Post-Transplant Complexity

David Park, age 61, presents for post-kidney transplant follow-up:

Transplant Medicine Assessment: "Patient 3 years post-kidney transplant with stable graft function. Current creatinine 1.4, stable from baseline. Managing chronic transplant nephropathy with slow decline in function over past year. Immunosuppression includes tacrolimus, mycophenolate, and prednisone with recent dose adjustments due to BK virus nephropathy."

Complication Management: "Recent surveillance biopsy showed grade 1A acute cellular rejection, now resolved with steroid pulse therapy. BK virus load detectable but declining with reduced immunosuppression. Managing post-transplant diabetes mellitus with insulin therapy."

Ongoing Monitoring: "Laboratory monitoring every 6 weeks for kidney function, drug levels, and infection surveillance. Annual surveillance biopsies continue. Patient education regarding infection prevention and medication compliance ongoing."

Comprehensive Coding:

- Z94.0 (Kidney transplant status) → HCC 176
- T86.10 (Unspecified complication of kidney transplant) → Additional complexity
- E11.9 (Type 2 diabetes mellitus without complications) → Post-transplant diabetes

The transplant status alone generates HCC 176 coefficient 0.367, adding approximately **$4,404** annually. Additional complications may qualify for supplementary HCC categories, reflecting the ongoing complexity of transplant care.

Rare Disease Recognition and Documentation

Rare diseases often generate substantial HCC payments due to their complexity, expensive treatments, and specialized care requirements. These conditions require precise documentation and may benefit from genetic testing results, specialty consultations, and treatment protocols.

Genetic Disorders including inherited metabolic conditions, chromosomal abnormalities, and genetic syndromes often require lifelong management and specialized treatments. Documentation should include genetic testing results when available and ongoing treatment requirements.

Orphan Diseases with few treatment options often involve experimental therapies, frequent hospitalizations, and intensive monitoring. These conditions justify enhanced payment due to their complexity and cost impact.

Multi-System Disorders affecting multiple organ systems require coordinated care across specialties and may qualify for multiple HCC categories. Documentation should capture all affected systems and treatment requirements.

Technology Integration for Complex Conditions

Advanced coding for respiratory, immune, and other system conditions benefits from technology tools that can identify complex clinical patterns, suggest appropriate code selections, and provide clinical decision support for challenging cases.

Clinical decision support systems can alert providers to potential drug interactions, monitoring requirements, and assessment needs specific to complex medical conditions. These systems enhance both patient safety and documentation quality.

Predictive analytics can identify patients at risk for complications or disease progression, enabling proactive intervention and documentation. These tools become particularly valuable for managing complex chronic conditions.

Integration with specialty systems including pulmonary function testing, immunology laboratories, and oncology treatment protocols can provide automated documentation support and coding suggestions based on clinical results.

Quality Assurance for Complex System Conditions

Complex medical conditions require robust quality assurance processes that ensure documentation adequately supports assigned codes while maintaining clinical accuracy. These processes become particularly important for high-value, less common conditions that may face audit scrutiny.

Clinical validation should include specialty consultation when available, ensuring that complex diagnoses reflect appropriate clinical expertise and diagnostic workup. This validation strengthens audit responses for unusual or high-value conditions.

Documentation review should address both clinical criteria and regulatory requirements, ensuring that complex conditions meet both medical necessity standards and HCC recognition criteria.

Provider education should address the specific documentation requirements for complex conditions, including staging criteria, treatment status indicators, and monitoring requirements that support ongoing HCC recognition.

Mastering System Complexity

Respiratory, immune, and other system conditions create substantial risk adjustment opportunities while requiring sophisticated understanding of diverse clinical presentations and documentation requirements. These conditions often involve expensive treatments, complex monitoring requirements, and multisystem effects that justify enhanced payment recognition.

Success in coding these complex conditions requires collaboration between clinical specialists, coding professionals, and quality assurance teams to ensure accurate capture while maintaining compliance standards. The diversity of these conditions demands broad clinical knowledge combined with specific expertise in documentation requirements and coding guidelines.

As treatment advances continue to improve outcomes for complex medical conditions, risk adjustment professionals must adapt their practices to reflect changing clinical realities while maintaining accurate payment recognition for legitimate clinical complexity.

Complex Systems Mastery Elements:

- COPD and respiratory conditions require evidence of active disease with ongoing treatment rather than historical diagnoses
- Autoimmune disorders involve expensive treatments and multisystem effects that justify enhanced payment recognition
- Cancer coding demands understanding of staging, treatment status, and the distinction between active disease and cancer history
- Transplant recipients require ongoing monitoring and immunosuppression that creates lifelong complexity requiring HCC recognition
- Rare diseases often generate substantial payments due to their complexity and specialized treatment requirements

Chapter 8: RADV Audit Mastery

Risk Adjustment Data Validation audits represent the most significant compliance challenge facing Medicare Advantage organizations today. These audits examine the foundation of risk adjustment payments by validating whether submitted diagnoses can be supported by adequate medical record documentation. With failure rates approaching 70% for some diagnosis categories and financial impacts reaching millions of dollars per audit, RADV mastery becomes essential for organizational survival and professional advancement.

The RADV audit process fundamentally changed with the implementation of the Final Rule creating new standards that require unprecedented levels of documentation quality and compliance rigor. Organizations that previously succeeded under traditional audit approaches now face substantial payment recoupments and compliance challenges that demand new strategies and enhanced expertise.

RADV Final Rule Implementation and Impact

The RADV Final Rule represents the most significant change in risk adjustment compliance since the inception of Medicare Advantage. This rule establishes new standards for documentation validation, introduces enhanced penalty structures, and creates audit methodologies that test organizational preparedness at unprecedented levels.

The Final Rule expands audit scope from 200 contracts annually to **550 contracts by 2025**, meaning virtually every Medicare Advantage organization will face regular audit scrutiny. This expansion creates a compliance environment where audit preparation becomes an ongoing operational requirement rather than an occasional concern.

Enhanced documentation standards under the Final Rule require evidence of active clinical management during the payment year for all submitted diagnoses. Historical documentation practices that satisfied previous audit requirements often fail under these enhanced standards, requiring comprehensive documentation improvement initiatives.

Statistical sampling methodologies create financial exposure that extends far beyond the audited sample size. Organizations may face payment adjustments affecting entire member populations based on error rates identified in relatively small audit samples, magnifying the financial impact of documentation deficiencies.

The rule establishes clear timelines for audit responses, limiting organizations' ability to gather additional documentation or clarify clinical circumstances after audit notification.

This constraint requires proactive documentation management and immediate audit response capabilities.

Case Example 1: RADV Audit Documentation Failure

Mountain View Health Plan received a RADV audit for member Barbara Johnson, age 73, with submitted diagnoses including major depressive disorder (HCC 155), chronic kidney disease stage 4 (HCC 326), and diabetes with complications (HCC 35).

Audit Request Documentation: The plan submitted three encounters from the payment year:

1. Primary care visit: "Depression stable on sertraline, kidney function monitored, diabetes well-controlled"
2. Endocrinology visit: "DM type 2, A1c 7.2%, continue metformin"
3. Nephrology visit: "CKD stage 4, creatinine 2.1, stable"

Audit Findings:

- **Major depressive disorder:** Mentioned in medication list but no clinical assessment, treatment adjustment, or symptom evaluation documented
- **Chronic kidney disease stage 4:** Lab value noted but no provider interpretation, staging confirmation, or treatment plan
- **Diabetes with complications:** No complications documented or evidence of complication management

Audit Results: All three diagnoses failed RADV validation, resulting in:

- Payment recoupment: **$24,600** for this member
- Extrapolated impact across member population: **$2.8 million**
- Enhanced compliance monitoring and potential enrollment sanctions

This case demonstrates how seemingly adequate documentation fails under RADV standards when it lacks evidence of active clinical management and treatment decision-making.

Common RADV Audit Findings and Failure Patterns

RADV audits reveal predictable failure patterns that organizations can address through targeted improvement initiatives. Understanding these patterns enables proactive remediation and improved audit outcomes for future validation efforts.

Insufficient Clinical Assessment represents the most common failure type, where medical records mention diagnoses without documenting clinical evaluation, assessment, or management decisions. These failures often involve conditions listed in problem lists or mentioned in passing without clinical context.

Inadequate Treatment Documentation occurs when records fail to demonstrate active treatment, monitoring, or management of submitted diagnoses. Auditors require evidence that providers actively addressed conditions rather than passively acknowledged their existence.

Missing MEAT Criteria failures involve diagnoses that lack evidence of monitoring, evaluation, assessment, or treatment during the payment year. These failures often affect chronic conditions that providers assume continue without documented clinical attention.

Historical Diagnosis Submissions create failures when organizations submit diagnoses based on historical documentation without evidence of current clinical relevance. These submissions often involve resolved conditions or complications that no longer require active management.

Provider Type Restrictions generate failures when diagnoses originate from providers who cannot generate risk adjustment diagnoses under CMS regulations. These failures often involve allied health professionals or providers operating outside their scope of practice.

Documentation Requirements for High-Risk Diagnoses

Certain diagnosis categories face enhanced audit scrutiny due to their high payment impact, frequent occurrence, or historical validation challenges. Understanding the specific documentation requirements for these conditions becomes essential for audit success.

Mental Health Conditions require evidence of psychiatric evaluation, treatment planning, and ongoing management rather than simple medication prescriptions or brief psychiatric references. Documentation should include symptom assessment, functional impact evaluation, and treatment response monitoring.

Heart Failure Diagnoses demand evidence of clinical assessment including symptom evaluation, physical examination findings, diagnostic testing interpretation, and treatment planning. Simple mentions of heart failure history without current clinical evaluation typically fail validation.

Diabetes Complications require clear documentation linking diabetic conditions to specific complications with evidence of ongoing management. Generic diabetes references without complication specification or management documentation often fail audit validation.

Kidney Disease Staging demands provider interpretation of laboratory values with clear staging designation and clinical correlation. Laboratory values alone without provider interpretation and staging confirmation typically fail RADV requirements.

Stroke and Neurological Conditions require documentation of current neurological deficits, functional impacts, and ongoing treatment rather than historical stroke references. Recent neurological assessments and functional evaluations strengthen validation outcomes.

Case Example 2: Successful RADV Documentation Strategy

Regional Health Network implemented comprehensive RADV preparation after experiencing significant audit failures:

Pre-Implementation Performance:

- RADV pass rate: 32%
- Average payment recoupment: $18,400 per failed member
- Common failure categories: Mental health (85% failure), heart failure (72% failure), diabetes complications (68% failure)

Documentation Enhancement Strategy:

1. Provider education focusing on MEAT criteria application
2. Electronic health record templates emphasizing clinical assessment
3. Real-time clinical decision support for high-risk diagnoses
4. Monthly audit simulation exercises using actual member records

Post-Implementation Documentation Example: Diabetes with Nephropathy Visit:
"Patient returns for diabetes management. Current A1c 7.8%, improved from 8.4% last visit *(Monitor)*. Physical exam shows no diabetic foot complications, blood pressure well-

controlled *(Evaluate)*. Laboratory results show creatinine 1.9, eGFR 32, confirming chronic kidney disease stage 3b secondary to diabetic nephropathy *(Assess)*. Continue ACE inhibitor for nephroprotection, increase metformin dose for glucose control, nephrology referral for CKD management *(Treatment)*."

Implementation Results (18 months):

- RADV pass rate improved to 87%
- Payment recoupment reduced to $3,200 per failed member
- Provider documentation scores increased 45%
- Estimated financial benefit: $12.3 million over two years

Audit Response Strategies and Timeline Management

Successful RADV audit responses require systematic approaches that maximize documentation quality while meeting strict timeline requirements. These strategies must balance thoroughness with efficiency to optimize audit outcomes within regulatory constraints.

Immediate Response Protocols should activate upon audit notification, including team assembly, record identification, and preliminary quality assessment. These protocols prevent delays that could compromise response quality or timeline compliance.

Documentation Gathering requires systematic collection of all relevant medical records from the audit period, including hospital records, specialty consultations, and diagnostic testing results. Comprehensive record gathering often identifies supporting documentation overlooked in initial chart reviews.

Clinical Review Processes should include clinical professionals who can interpret medical documentation and identify the strongest supporting evidence for each audited diagnosis. These reviews ensure that audit responses emphasize the most compelling clinical evidence.

Response Preparation involves organizing documentation in clear, logical presentations that highlight MEAT criteria evidence and clinical reasoning. Effective responses guide auditors through the clinical narrative rather than requiring them to search for relevant information.

Quality Assurance reviews should occur before submission to identify potential weaknesses and ensure that responses address all audit requirements. These reviews often identify additional supporting evidence or presentation improvements that strengthen audit outcomes.

Appeals and Rebuttal Process Mastery

Organizations that receive adverse RADV audit findings have appeal rights that can result in overturned decisions and payment restoration. Understanding the appeals process and developing effective rebuttal strategies becomes essential for protecting organizational interests.

Appeal Timeline Requirements allow limited time periods for filing appeals and providing additional documentation. Organizations must maintain strict timeline compliance to preserve their appeal rights and avoid automatic adverse determinations.

Additional Documentation may be submitted during appeals when it was available during the original audit period but not initially provided. This opportunity allows organizations to strengthen their cases with comprehensive supporting evidence.

Clinical Expert Review often proves valuable during appeals by providing independent clinical interpretation of medical documentation. Expert reviews can clarify clinical reasoning and demonstrate appropriate care standards that support diagnosis validity.

Legal Consultation may be appropriate for appeals involving significant financial exposure or complex regulatory interpretation. Legal experts can provide guidance regarding procedural requirements and substantive arguments that strengthen appeal positions.

Case Example 3: Successful RADV Appeal Strategy

Central Coast Health Plan faced a $4.2 million payment recoupment after RADV audit failures across multiple members:

Initial Audit Results:

- 47 audited diagnoses failed validation
- Primary failure categories: Heart failure, diabetes complications, mental health conditions
- Auditor rationale: Insufficient clinical assessment and treatment documentation

Appeal Strategy Development:

1. Clinical expert review of all failed diagnoses
2. Comprehensive documentation gathering including hospital records and consultant notes
3. Timeline development of clinical care progression
4. Focus on clinical reasoning and care standards

Appeal Documentation Enhancement: For member with failed heart failure diagnosis, the appeal included:

- Echocardiogram reports showing reduced ejection fraction
- Cardiology consultation notes with treatment recommendations
- Emergency department records showing heart failure exacerbations
- Hospital discharge summaries with heart failure management plans
- Pharmacy records showing heart failure medication dispensing

Appeal Outcome:

- 31 of 47 failed diagnoses were overturned on appeal
- Payment recoupment reduced from $4.2 million to $1.6 million
- Lessons learned applied to prevent future audit failures
- Enhanced documentation protocols implemented across provider network

Internal Audit Programs and Preparation

Proactive internal audit programs provide early identification of RADV vulnerabilities while building organizational capabilities for external audit success. These programs should simulate actual RADV methodologies to provide realistic assessments of audit readiness.

Risk Stratification should identify members and diagnoses with highest audit risk based on payment impact, clinical complexity, and documentation quality indicators. This stratification focuses internal audit resources on areas most likely to face external scrutiny.

Sampling Methodologies should mirror CMS RADV sampling approaches to provide realistic assessments of organizational audit exposure. These methodologies help organizations understand their potential financial exposure and compliance risks.

Documentation Review Criteria should apply actual RADV validation standards to provide accurate assessments of audit readiness. These criteria must reflect current CMS guidance and established audit precedents.

Provider Feedback should include specific, actionable recommendations for documentation improvement based on internal audit findings. This feedback helps providers understand requirements while building documentation capabilities.

Trending Analysis should identify patterns in documentation deficiencies and track improvement over time. This analysis enables targeted interventions and demonstrates progress in audit readiness.

Technology Solutions for RADV Compliance

Advanced technology solutions can enhance RADV compliance by automating documentation review, identifying potential vulnerabilities, and providing real-time feedback regarding audit readiness. These solutions must balance automation with clinical judgment to ensure accuracy and effectiveness.

Automated Documentation Review can scan medical records to identify MEAT criteria evidence and flag potential documentation deficiencies. These systems require careful calibration to minimize false positives while capturing genuine risks.

Predictive Analytics can identify members and diagnoses at highest risk for RADV audit failure based on documentation patterns and clinical characteristics. These analytics enable proactive intervention before audit exposure occurs.

Clinical Decision Support can provide real-time feedback during clinical encounters regarding documentation requirements and potential audit risks. These systems integrate with electronic health records to enhance provider workflow.

Audit Simulation Tools can regularly assess organizational RADV readiness using actual CMS criteria and provide ongoing monitoring of compliance trends. These tools enable continuous improvement and early risk identification.

Conquering the RADV Challenge

RADV audit mastery requires comprehensive understanding of regulatory requirements, documentation standards, audit processes, and response strategies. The stakes continue to rise as audit frequency increases and financial impacts grow, making RADV expertise essential for organizational success and professional advancement.

Success in RADV management requires collaboration between clinical providers, coding professionals, compliance teams, and organizational leadership to create comprehensive strategies that address both immediate audit challenges and long-term compliance sustainability. The organizations that master these challenges will thrive in the evolving risk adjustment environment.

The RADV audit environment will continue to evolve as CMS refines its approaches and organizations adapt their practices. Professionals who develop deep expertise in RADV compliance will find themselves essential to organizational success and well-positioned for career advancement in the growing risk adjustment field.

RADV Mastery Framework:

- RADV Final Rule implementation creates unprecedented compliance challenges requiring enhanced documentation and audit preparation
- Common audit failure patterns include insufficient clinical assessment, inadequate treatment documentation, and missing MEAT criteria
- High-risk diagnoses require specific documentation approaches that demonstrate active clinical management and treatment
- Successful audit responses demand systematic approaches that maximize documentation quality within strict timeline requirements
- Internal audit programs and technology solutions provide proactive compliance management and early risk identification

Chapter 9: Compliance and Ethics

Risk adjustment compliance encompasses far more than accurate coding and documentation—it requires understanding complex regulatory frameworks, recognizing potential fraud risks, and implementing ethical practices that protect both organizations and patients. The intersection of massive financial incentives, subjective clinical judgments, and evolving regulatory requirements creates an environment where compliance expertise becomes essential for both individual success and organizational survival.

The Office of Inspector General, Department of Justice, and Centers for Medicare & Medicaid Services maintain intense focus on risk adjustment practices, with enforcement actions resulting in hundreds of millions of dollars in penalties and settlements. Understanding these enforcement priorities and implementing robust compliance programs becomes essential for managing regulatory risk while optimizing legitimate risk adjustment opportunities.

OIG Focus Areas and Enforcement Trends

The Office of Inspector General has identified risk adjustment as a priority area for investigation and enforcement, publishing specific guidance regarding acceptable practices and common violation patterns. Understanding these focus areas enables organizations to align their practices with regulatory expectations while avoiding problematic approaches that invite scrutiny.

Upcoding Investigations represent the most common OIG enforcement action, where organizations allegedly submit diagnosis codes that are more severe or complex than supported by medical record documentation. These investigations often involve statistical analysis of coding patterns compared to industry benchmarks and clinical expectations.

Unsupported Diagnosis Submissions create enforcement risk when organizations submit diagnoses without adequate medical record support or clinical justification. These violations often involve diagnoses that generate high payments but lack corresponding clinical evidence or treatment requirements.

Provider Inducement investigations examine financial arrangements between health plans and providers that may inappropriately influence coding and documentation practices. These arrangements require careful structuring to ensure compliance with anti-kickback and inducement regulations.

Documentation Manipulation creates significant enforcement risk when organizations or providers create or modify medical record documentation specifically to support risk adjustment submissions. These practices constitute fraud regardless of the clinical accuracy of submitted diagnoses.

The OIG's enforcement approach increasingly relies on sophisticated data analytics that identify statistical anomalies and coding patterns suggesting potential violations. Organizations must understand these analytical approaches and ensure their practices can withstand statistical scrutiny.

Case Example 1: OIG Settlement Analysis

Western Regional Health Plan faced a $47 million settlement with the OIG following a comprehensive risk adjustment investigation:

Investigation Findings:

- Risk adjustment diagnoses increased 340% over three years without corresponding medical complexity increases
- 23% of submitted diagnoses lacked adequate medical record support
- Provider contracts included productivity bonuses potentially tied to HCC capture
- Internal communications suggested pressure on providers to increase diagnosis submissions

Specific Violations Identified:

1. Submission of diabetes complication codes without evidence of active complications
2. Mental health diagnoses supported only by medication lists without clinical assessment
3. Chronic kidney disease staging without laboratory interpretation or provider confirmation
4. Historical diagnosis submissions without evidence of current clinical relevance

Settlement Terms:

- $47 million financial penalty
- 5-year corporate integrity agreement with ongoing monitoring
- Required compliance program enhancements including annual audits
- Provider training requirements and certification processes

- Enhanced documentation standards and review procedures

Lessons for Industry: This case demonstrates how aggressive risk adjustment practices without adequate compliance safeguards can result in massive financial penalties and operational restrictions. The settlement emphasizes the importance of documentation quality, provider education, and robust compliance monitoring.

False Claims Act Implications for Risk Adjustment

The False Claims Act creates significant liability for organizations that submit inaccurate or unsupported risk adjustment diagnoses. Understanding FCA requirements and potential violations becomes essential for developing compliant risk adjustment practices and avoiding devastating legal exposure.

Knowingly Submitting False Claims includes actual knowledge, deliberate ignorance, or reckless disregard regarding claim accuracy. Organizations cannot avoid liability by claiming ignorance of documentation deficiencies or coding errors when they should have known about potential problems.

Materiality Standards require that false claims influence government payment decisions to create FCA liability. Given the substantial payment implications of risk adjustment diagnoses, virtually all unsupported submissions likely meet materiality thresholds for FCA violations.

Individual Liability extends beyond organizational responsibility to include individual employees who participate in or facilitate false claim submissions. This personal liability creates strong incentives for compliance professionals to implement robust safeguards and quality assurance processes.

Whistleblower Protections encourage reporting of potential violations by current and former employees, creating internal compliance risks for organizations with inadequate training, oversight, or ethical cultures. These protections include financial rewards for successful whistleblower actions.

Qui Tam Actions allow private individuals to bring FCA lawsuits on behalf of the government, creating additional enforcement mechanisms beyond traditional government investigations. These actions often involve former employees or competitors with detailed knowledge of organizational practices.

Ethical Dilemmas in Risk Adjustment Practice

Risk adjustment professionals frequently encounter ethical situations that require careful analysis and principled decision-making. Understanding common ethical dilemmas and developing frameworks for resolution becomes essential for maintaining professional integrity while achieving organizational objectives.

Documentation Improvement versus Documentation Creation represents a fundamental ethical distinction in risk adjustment practice. Legitimate documentation improvement helps providers create complete and accurate records, while documentation creation involves generating diagnoses or clinical information not supported by actual patient encounters.

Aggressive Coding versus Accurate Coding requires balancing organizational financial incentives with professional coding standards and regulatory requirements. Ethical practitioners focus on capturing all appropriate diagnoses while avoiding unsupported or inflated coding practices.

Provider Education versus Provider Inducement distinguishes between legitimate educational efforts that improve clinical documentation and inappropriate inducements that encourage providers to submit unsupported diagnoses for financial benefit.

Quality versus Quantity in risk adjustment capture requires prioritizing accuracy and compliance over simple volume metrics. Ethical programs emphasize sustainable practices that support long-term compliance rather than short-term revenue maximization.

Patient Privacy versus Audit Requirements creates tension between protecting patient confidentiality and meeting regulatory documentation requirements. Compliance programs must balance these competing obligations while maintaining ethical standards.

Case Example 2: Ethical Decision-Making Framework

Sarah Chen, CRC, faces an ethical dilemma when reviewing medical records for risk adjustment submission:

Situation: A primary care provider's documentation states: "Patient reports previous heart attack several years ago, currently asymptomatic and not requiring cardiac medications. Will monitor."

Ethical Analysis:

- **Clinical Reality:** Patient has history of myocardial infarction but no current cardiac treatment or active management
- **Coding Options:** Could assign I25.2 (Old myocardial infarction) mapping to HCC 160 with high coefficient
- **Documentation Quality:** Mentions historical condition but lacks evidence of current clinical relevance
- **MEAT Criteria:** No evidence of current monitoring, evaluation, assessment, or treatment

Decision Framework:

1. **Professional Standards:** CRC code of ethics requires accuracy and integrity
2. **Regulatory Requirements:** RADV audits require evidence of active management
3. **Clinical Appropriateness:** Historical condition without current relevance questionable
4. **Organizational Impact:** High-value HCC with substantial payment implications

Ethical Resolution: Sarah recommends not submitting the diagnosis due to inadequate documentation of current clinical management. She provides feedback to the clinical documentation improvement team regarding proper documentation of historical cardiac conditions requiring ongoing monitoring.

Outcome: The organization implements enhanced provider education regarding documentation of historical conditions with current clinical significance, improving both compliance and ethical practices.

Internal Audit Best Practices and Program Development

Effective compliance programs require robust internal audit capabilities that identify potential problems before they become regulatory violations. These programs must balance thoroughness with efficiency while maintaining independence and objectivity.

Risk Assessment should identify areas of highest compliance risk based on regulatory focus, financial impact, and organizational practices. This assessment guides audit resource allocation and priority setting for compliance activities.

Audit Planning should include both scheduled reviews and responsive audits triggered by specific indicators or concerns. Planned audits provide ongoing monitoring while responsive audits address emerging issues or identified problems.

Sampling Methodologies should provide statistically valid assessments of compliance performance while focusing on high-risk areas and practices. These methodologies must balance coverage with resource constraints.

Documentation Standards for internal audits should mirror external audit requirements to provide realistic assessments of regulatory compliance. These standards ensure that internal audits accurately predict external audit outcomes.

Corrective Action Processes should address identified deficiencies through systematic remediation that prevents recurrence. These processes must include accountability mechanisms and progress monitoring.

Independence Requirements ensure that internal audit functions maintain objectivity and credibility. These requirements typically involve reporting relationships that separate audit functions from operational management.

Compliance Program Elements and Structure

Comprehensive compliance programs integrate multiple elements that work together to prevent violations, detect problems, and respond appropriately to identified issues. Understanding these elements and their interactions becomes essential for developing effective compliance frameworks.

Written Policies and Procedures provide the foundation for compliance programs by establishing clear standards, expectations, and requirements. These documents must address specific risk adjustment practices and regulatory requirements.

Training and Education ensure that all personnel understand compliance requirements and their individual responsibilities. These programs must address role-specific risks and provide ongoing reinforcement of compliance principles.

Monitoring and Auditing provide ongoing assessment of compliance performance and early identification of potential problems. These activities must be sufficiently robust to detect violations before they become systemic issues.

Response and Remediation processes ensure appropriate handling of identified violations and implementation of corrective measures. These processes must include investigation capabilities, corrective action development, and progress monitoring.

Enforcement and Discipline mechanisms provide accountability for compliance violations and reinforce the importance of ethical behavior. These mechanisms must be consistent, fair, and proportionate to violation severity.

Communication and Reporting systems enable identification and escalation of compliance concerns while protecting individuals who report potential violations. These systems must include confidential reporting mechanisms and whistleblower protections.

Case Example 3: Comprehensive Compliance Program Implementation

Metro Health Solutions implemented a comprehensive risk adjustment compliance program following industry enforcement trends:

Program Structure:

- Chief Compliance Officer reporting directly to CEO and Board
- Dedicated risk adjustment compliance team with clinical and coding expertise
- Monthly compliance committee meetings with cross-functional participation
- Annual compliance risk assessments with board-level review

Key Program Elements:

1. Policies and Procedures:

- Risk adjustment coding standards aligned with CMS guidance
- Documentation requirements exceeding minimum regulatory standards
- Provider contract language addressing compliance expectations
- Clear escalation procedures for compliance concerns

2. Training and Education:

- Annual compliance training for all risk adjustment personnel
- Role-specific training addressing individual responsibilities
- Provider education programs focusing on documentation requirements
- Regular updates addressing regulatory changes and enforcement trends

3. Monitoring and Auditing:

- Monthly internal audits using RADV criteria

- Annual third-party compliance assessments
- Real-time monitoring of coding accuracy and documentation quality
- Statistical analysis of risk adjustment trends and patterns

4. Investigation and Response:

- Defined investigation procedures for compliance concerns
- Corrective action protocols addressing identified deficiencies
- Self-disclosure procedures for potential violations
- Coordination with legal counsel for significant issues

Implementation Results (24 months):

- Internal audit pass rates improved from 78% to 94%
- Compliance training completion rates: 99.7%
- Zero regulatory enforcement actions or investigations
- Enhanced provider satisfaction with documentation support
- Estimated compliance program ROI: 12:1 based on avoided penalties

Emerging Compliance Challenges and Future Trends

The risk adjustment compliance environment continues to evolve as regulators develop new enforcement approaches and industry practices advance. Understanding emerging trends enables proactive compliance management and preparation for future regulatory developments.

Artificial Intelligence and Automation create new compliance considerations as organizations increasingly rely on technology for coding assistance and documentation review. These technologies require careful oversight to ensure they support rather than compromise compliance objectives.

Data Analytics and Pattern Recognition enable both improved compliance monitoring and enhanced regulatory enforcement. Organizations must develop analytical capabilities while ensuring their practices can withstand sophisticated statistical scrutiny.

Social Determinants Integration may create new compliance considerations as risk adjustment models evolve to address non-medical factors affecting healthcare costs. These developments require careful navigation of privacy, accuracy, and appropriate use principles.

Value-Based Care Integration creates additional complexity as organizations balance risk adjustment optimization with quality measure performance and patient outcome improvement. These competing priorities require sophisticated management approaches.

Provider Network Expansion creates compliance challenges as organizations work with diverse provider types and practice patterns. Maintaining consistent compliance standards across varied provider relationships requires enhanced management capabilities.

Building Ethical Excellence in Risk Adjustment

Risk adjustment compliance extends beyond regulatory requirement adherence to encompass ethical practice standards that protect patients, providers, and organizations. Building ethical excellence requires cultural commitment, systematic processes, and ongoing reinforcement of professional standards.

Ethical Leadership sets the tone for organizational compliance by demonstrating commitment to integrity and appropriate practices. This leadership must be visible, consistent, and supported by concrete actions and resource allocation.

Professional Development should include ethics training that addresses real-world dilemmas and provides frameworks for principled decision-making. This development helps professionals navigate complex situations while maintaining ethical standards.

Industry Engagement through professional organizations and peer networks provides access to best practices and emerging compliance trends. This engagement enables learning from industry experience while contributing to professional standards development.

Continuous Improvement approaches recognize that compliance excellence requires ongoing refinement and adaptation to changing circumstances. These approaches emphasize learning, growth, and proactive enhancement of compliance capabilities.

Safeguarding Integrity in High-Stakes Environment

Risk adjustment compliance operates in an environment where massive financial incentives intersect with complex regulatory requirements and subjective clinical judgments. Success requires more than technical knowledge—it demands ethical commitment, systematic processes, and unwavering focus on integrity and appropriate practices.

The professionals who master compliance excellence while maintaining ethical standards will find themselves essential to organizational success and well-positioned for leadership roles in the evolving healthcare environment. This mastery requires ongoing investment in knowledge, skills, and professional development combined with unwavering commitment to doing the right thing.

Compliance excellence serves multiple stakeholders: protecting organizations from regulatory risk, ensuring appropriate payment for legitimate clinical complexity, and maintaining public trust in risk adjustment programs. This broad impact makes compliance expertise both personally rewarding and professionally essential.

Compliance Excellence Framework:

- OIG enforcement focuses on upcoding, unsupported diagnoses, and inappropriate provider arrangements requiring proactive compliance management
- False Claims Act liability extends to individuals and organizations submitting inaccurate risk adjustment diagnoses with severe financial and legal consequences
- Ethical dilemmas require principled decision-making frameworks that balance competing interests while maintaining professional integrity
- Internal audit programs provide early identification of compliance risks and validation of program effectiveness
- Comprehensive compliance programs integrate multiple elements including policies, training, monitoring, and response capabilities

Chapter 10: Data Analytics and Predictive Modeling

Risk adjustment success increasingly depends on sophisticated analytical capabilities that transform raw data into actionable insights for clinical, operational, and financial decision-making. The intersection of massive healthcare datasets, advanced analytical techniques, and artificial intelligence creates unprecedented opportunities for organizations that can harness these tools effectively. Understanding predictive modeling fundamentals, implementing analytical frameworks, and leveraging technology solutions becomes essential for competitive advantage in the evolving risk adjustment environment.

Modern risk adjustment programs generate enormous amounts of data from clinical encounters, coding activities, and member health status changes. Organizations that can analyze these data streams to identify patterns, predict outcomes, and optimize interventions will outperform competitors relying solely on traditional coding approaches. This analytical transformation requires new skills, technologies, and organizational capabilities that extend far beyond conventional medical coding expertise.

Predictive Modeling Fundamentals in Healthcare

Predictive modeling in risk adjustment uses historical data patterns to forecast future health events, identify coding opportunities, and predict financial outcomes. These models combine clinical indicators, utilization patterns, demographic factors, and social determinants to create comprehensive risk profiles that guide intervention strategies and resource allocation decisions.

Regression Analysis forms the foundation of most risk adjustment predictive models, identifying relationships between independent variables (clinical conditions, demographics, utilization) and dependent variables (future costs, health outcomes, coding accuracy). These relationships enable prediction of member risk scores, expected healthcare utilization, and potential coding gaps.

Machine Learning Algorithms including random forests, gradient boosting, and neural networks can identify complex patterns in healthcare data that traditional statistical methods might miss. These algorithms excel at handling large datasets with multiple variables and can adapt to changing patterns over time.

Time Series Analysis addresses the temporal nature of healthcare data, recognizing that patient conditions, utilization patterns, and coding accuracy change over time. These models account for seasonal variations, disease progression, and treatment response patterns that affect risk adjustment outcomes.

Survival Analysis predicts time-to-event outcomes such as hospital admissions, emergency department visits, or disease progression. These models help identify high-risk members who may benefit from proactive interventions or enhanced case management.

Predictive models require careful validation to ensure they perform accurately on new data and generalize beyond the training dataset. Cross-validation, holdout testing, and temporal validation help ensure model reliability and prevent overfitting that could lead to poor real-world performance.

Case Example 1: Predictive Model for Heart Failure Identification

Regional Medical Group developed a predictive model to identify Medicare Advantage members likely to have undiagnosed or undocumented heart failure:

Model Development Process:

1. **Data Collection:** Gathered 3 years of claims data, laboratory results, medication histories, and coded diagnoses for 45,000 members
2. **Feature Engineering:** Created variables including diuretic prescriptions, cardiology visits, echocardiogram results, BNP levels, and comorbidity patterns
3. **Model Training:** Used gradient boosting algorithm with 70% training, 15% validation, 15% testing split
4. **Validation:** Achieved 87% sensitivity and 82% specificity in identifying members with heart failure

Key Predictive Variables:

- Loop diuretic prescriptions (strongest predictor)
- Multiple cardiology visits within 6 months
- Elevated BNP or NT-proBNP results
- Combination of diabetes, hypertension, and coronary disease
- Emergency department visits with respiratory complaints

Implementation Results:

- Identified 2,847 members with high probability of heart failure

- Clinical review confirmed heart failure in 2,204 members (77% accuracy)
- Additional HCC capture generated $7.2 million in annual payments
- Enhanced care management reduced hospitalizations by 23%

Model Refinement: Quarterly model updates incorporated new data and feedback from clinical reviews, improving accuracy to 89% sensitivity and 85% specificity within 18 months.

Analytics for Identifying Coding Gaps and Opportunities

Data analytics can systematically identify members with potential coding gaps by analyzing patterns in clinical data, utilization, medications, and existing diagnoses. These analytical approaches enable targeted chart reviews and provider outreach that maximize coding accuracy while minimizing unnecessary administrative burden.

Pharmaceutical Analytics examine medication patterns to identify potential diagnoses not captured in medical coding. For example, members receiving insulin without diabetes diagnoses, or those taking antipsychotics without documented mental health conditions. These analyses require understanding of medication indications and contraindications.

Utilization Pattern Analysis identifies members whose healthcare utilization suggests undiagnosed or undocumented conditions. High emergency department utilization might indicate poorly controlled chronic conditions, while frequent specialty visits could suggest complex conditions not reflected in primary care coding.

Laboratory Value Analysis can identify clinical conditions suggested by abnormal test results but not captured in diagnosis coding. Elevated creatinine levels might indicate chronic kidney disease, while abnormal HbA1c values could suggest undiagnosed diabetes or poor diabetes control.

Provider Specialty Patterns analyze which specialists members visit to infer potential diagnoses. Oncology visits suggest cancer diagnoses, while endocrinology visits might indicate diabetes, thyroid disorders, or other endocrine conditions requiring capture.

Comorbidity Clustering identifies members with multiple related conditions who might have additional diagnoses based on common disease patterns. Members with diabetes and hypertension have high probability of cardiovascular disease, while those with COPD often develop heart failure.

Suspicious Pattern Detection and Fraud Prevention

Advanced analytics can identify potentially problematic coding patterns that might indicate fraud, abuse, or inappropriate practices. These analytical approaches help organizations maintain compliance while protecting against regulatory scrutiny and financial penalties.

Statistical Outlier Detection identifies providers, members, or diagnosis patterns that deviate significantly from expected norms. Providers with unusually high rates of specific high-value diagnoses might require additional review and education.

Temporal Pattern Analysis examines diagnosis timing patterns to identify potential manipulation. For example, large increases in diagnosis submissions near year-end might suggest inappropriate timing of diagnosis capture rather than legitimate clinical identification.

Documentation Quality Scoring uses natural language processing to assess medical record quality and identify potential documentation deficiencies before they become audit problems. These scores can guide targeted documentation improvement efforts.

Provider Benchmarking compares individual provider coding patterns against peer groups to identify outliers requiring additional attention. This benchmarking must account for patient population differences and practice patterns.

Cross-Reference Validation compares diagnosis submissions against supporting clinical data to identify potential inconsistencies. Claims for diabetes medications without diabetes diagnoses, or mental health diagnoses without corresponding treatment patterns, might indicate coding errors.

Case Example 2: Fraud Detection Analytics Implementation

Metro Health Plan implemented comprehensive fraud detection analytics after experiencing audit challenges:

Analytics Framework Development:

1. **Provider Risk Scoring:** Developed algorithms scoring providers based on coding patterns, diagnosis timing, and peer comparisons
2. **Member Profile Analysis:** Created models identifying members with potentially inflated risk scores based on utilization patterns

3. **Documentation Quality Assessment:** Implemented natural language processing to evaluate clinical note quality

Suspicious Pattern Identification:

- Provider A: 340% increase in depression diagnoses over 6 months without corresponding mental health treatment
- Provider B: Chronic kidney disease submissions without supporting laboratory values or nephrology referrals
- Provider C: Heart failure diagnoses concentrated in December submissions with limited prior cardiac care

Investigation and Resolution:

1. **Clinical Review:** Internal team reviewed medical records for flagged cases
2. **Provider Education:** Conducted targeted training for providers with concerning patterns
3. **Policy Enhancement:** Implemented enhanced documentation requirements for high-risk diagnoses
4. **Ongoing Monitoring:** Established quarterly analytics reviews with provider feedback

Program Results:

- Identified and corrected $3.8 million in inappropriate risk adjustment submissions
- Reduced audit risk through proactive compliance management
- Improved provider documentation quality through targeted education
- Enhanced organizational reputation for compliance excellence

Technology Integration Strategies and Implementation

Successful analytics programs require sophisticated technology infrastructure that can handle large healthcare datasets while providing user-friendly interfaces for clinical and operational teams. These implementations must balance analytical power with practical usability.

Data Warehouse Development creates centralized repositories that integrate data from multiple sources including electronic health records, claims systems, laboratory results, and pharmacy data. These warehouses enable comprehensive analysis while maintaining data quality and security.

Real-Time Analytics Platforms provide immediate feedback regarding coding opportunities, documentation quality, and potential compliance risks. These platforms integrate with clinical workflows to provide actionable insights without disrupting provider productivity.

Machine Learning Infrastructure supports model development, training, and deployment while ensuring scalability and performance. These infrastructures must handle both batch processing for large-scale analysis and real-time scoring for individual decisions.

Visualization and Reporting Tools translate complex analytical results into understandable formats for clinical and operational teams. These tools must provide both high-level dashboards and detailed drill-down capabilities.

Integration with Clinical Systems ensures that analytical insights reach decision-makers at the point of care. These integrations might include clinical decision support alerts, documentation templates, or care management notifications.

Artificial Intelligence and Machine Learning Applications

AI and machine learning applications in risk adjustment continue to advance rapidly, creating new opportunities for improved accuracy, efficiency, and clinical outcomes. Understanding these applications and their limitations becomes essential for organizations considering advanced technology investments.

Natural Language Processing can extract diagnostic information from unstructured clinical notes, identifying potential HCC diagnoses not captured in structured coding. These systems require careful training and validation to ensure accuracy and clinical relevance.

Computer Vision applications can analyze medical images to identify conditions relevant to risk adjustment. Chest X-rays might reveal evidence of heart failure or COPD, while fundoscopic images could identify diabetic retinopathy.

Predictive Coding systems can suggest potential diagnoses based on clinical patterns, medications, and utilization data. These systems require human oversight to ensure clinical appropriateness and regulatory compliance.

Automated Documentation Review can assess medical record quality and identify potential audit risks before submission. These systems can flag missing MEAT criteria evidence or insufficient clinical support for specific diagnoses.

Risk Score Optimization algorithms can identify the most effective interventions for improving member risk scores while maintaining clinical appropriateness and compliance standards.

Case Example 3: AI-Powered Clinical Decision Support

Valley Health Network implemented an AI-powered clinical decision support system for risk adjustment optimization:

System Components:

1. **NLP Engine:** Analyzed clinical notes to identify potential HCC diagnoses mentioned but not coded
2. **Predictive Algorithms:** Identified members likely to have specific conditions based on clinical patterns
3. **Documentation Assistant:** Provided real-time suggestions for improving clinical documentation
4. **Quality Monitor:** Assessed documentation quality using RADV audit criteria

Clinical Integration:

- Real-time alerts during clinical encounters suggesting potential diagnoses
- Documentation templates emphasizing MEAT criteria completion
- Provider dashboards showing coding accuracy and improvement opportunities
- Monthly reports highlighting successful interventions and outcomes

Implementation Results (18 months):

- Risk score accuracy improved 15% with enhanced clinical validity
- Provider documentation quality scores increased 28%
- RADV audit simulation pass rates improved from 72% to 91%
- Provider satisfaction with technology support increased significantly
- Estimated annual financial benefit: $12.4 million

Lessons Learned:

- AI systems require extensive training and ongoing calibration
- Provider acceptance depends on workflow integration and clear value demonstration
- Human oversight remains essential for clinical appropriateness and compliance
- Success requires organizational commitment to change management and training

Population Health Applications and Outcomes

Risk adjustment analytics extend beyond coding optimization to support broader population health management goals including care coordination, disease management, and outcome improvement. These applications demonstrate the clinical value of risk adjustment programs while supporting financial objectives.

Risk Stratification uses analytical models to categorize members based on health status, utilization patterns, and intervention needs. These stratifications guide resource allocation and care management program design.

Care Gap Identification combines clinical guidelines with member data to identify opportunities for preventive care, chronic disease management, and quality improvement interventions.

Intervention Targeting uses predictive models to identify members most likely to benefit from specific interventions such as disease management programs, care coordination, or medication management.

Outcome Prediction forecasts health outcomes, costs, and utilization patterns to support care planning and resource allocation decisions. These predictions enable proactive intervention and improved care coordination.

Program Evaluation measures the effectiveness of population health interventions through analytical comparison of intervention and control groups. These evaluations guide program refinement and resource allocation decisions.

Building Analytical Capabilities and Organizational Competency

Developing sophisticated analytical capabilities requires organizational investment in technology, personnel, and processes that support data-driven decision making. These capabilities must align with organizational objectives while maintaining focus on clinical quality and regulatory compliance.

Data Governance establishes policies and procedures for data quality, security, and appropriate use. These frameworks ensure that analytical initiatives comply with privacy regulations while maintaining data integrity.

Analytical Team Development requires recruiting personnel with appropriate statistical, clinical, and technology skills. These teams must combine technical expertise with healthcare knowledge to develop meaningful and actionable insights.

Technology Infrastructure must support both current analytical needs and future growth while maintaining security and performance standards. These infrastructures require ongoing investment and maintenance.

Change Management processes help organizations adapt to analytical insights and modify practices based on data-driven recommendations. These processes must address both technical and cultural barriers to change.

Performance Measurement tracks the effectiveness of analytical initiatives through defined metrics and regular evaluation. These measurements guide investment decisions and program refinement.

Transforming Data into Strategic Advantage

Analytics and predictive modeling represent the future of risk adjustment management, transforming how organizations identify opportunities, manage compliance risks, and optimize clinical and financial outcomes. The organizations that master these capabilities will achieve sustainable competitive advantages while improving patient care and organizational performance.

Success in analytics requires more than technology implementation—it demands organizational commitment to data-driven decision making, investment in appropriate capabilities, and cultural adaptation to analytical insights. The professionals who develop expertise in healthcare analytics while maintaining clinical focus will find themselves essential to organizational success.

The analytical transformation of risk adjustment continues to accelerate as technology capabilities advance and data sources expand. Professionals who master these tools while maintaining focus on clinical quality and regulatory compliance will shape the future of risk adjustment and healthcare delivery.

Analytics Excellence Framework:

- Predictive modeling fundamentals enable forecasting of health events, coding opportunities, and financial outcomes using sophisticated statistical techniques
- Coding gap identification through pharmaceutical, utilization, and laboratory analytics maximizes capture while maintaining clinical appropriateness
- Suspicious pattern detection protects organizations from fraud risks while ensuring compliance with regulatory requirements
- Technology integration requires sophisticated infrastructure that balances analytical power with practical usability for clinical teams
- AI and machine learning applications create new opportunities for accuracy and efficiency while requiring careful oversight and validation

Chapter 11: Quality Measures and Star Ratings

Risk adjustment and quality measurement intersect in complex ways that affect both financial performance and regulatory compliance for Medicare Advantage organizations. Understanding these relationships enables strategic alignment of coding activities with quality improvement initiatives while maximizing both payment optimization and member outcomes. The Star Ratings program creates additional financial incentives that can dwarf risk adjustment impacts, making quality-risk adjustment integration essential for organizational success.

Quality measures often depend on the same clinical conditions captured through risk adjustment coding, creating opportunities for coordinated improvement initiatives that address both payment accuracy and quality performance. However, these measures also create tension between aggressive coding practices and conservative clinical management, requiring careful balance to achieve optimal outcomes across multiple performance domains.

HEDIS Measure Integration with Risk Adjustment

Healthcare Effectiveness Data and Information Set measures create direct connections between risk adjustment coding and quality performance in areas such as diabetes management, cardiovascular care, and behavioral health treatment. Understanding these connections enables coordinated strategies that improve both payment accuracy and quality outcomes.

Comprehensive Diabetes Care measures require identification of diabetic members through claims data or encounter diagnoses, making accurate diabetes coding essential for both HCC capture and quality measure performance. Organizations must ensure that diabetes diagnoses submitted for risk adjustment also support quality measure denominators.

Controlling High Blood Pressure measures depend on hypertension diagnosis identification, creating alignment between HCC capture for cardiovascular conditions and quality measure participation. Accurate hypertension coding supports both payment optimization and quality performance measurement.

Antidepressant Medication Management requires identification of members with depression diagnoses, aligning with mental health HCC categories. Organizations can coordinate depression identification efforts to support both risk adjustment and quality measure performance.

Medication Adherence measures across multiple therapeutic categories benefit from accurate diagnosis coding that identifies eligible members and supports intervention targeting. Risk adjustment coding activities can feed quality measure identification and management programs.

Breast Cancer Screening and other preventive care measures create opportunities for diagnosis identification during screening encounters. Organizations can use quality measure activities to identify previously unknown conditions that qualify for risk adjustment recognition.

Case Example 1: Integrated Diabetes Management Program

Coastal Health Plan developed an integrated program addressing both diabetes risk adjustment and HEDIS performance:

Program Design:

1. **Member Identification:** Used both risk adjustment coding and pharmacy data to identify diabetic members
2. **Care Gap Analysis:** Identified members missing diabetes complications documentation and HEDIS care requirements
3. **Provider Outreach:** Coordinated education addressing both documentation improvement and quality care delivery
4. **Technology Integration:** Enhanced EHR alerts for both coding opportunities and quality measures

Diabetes Care Coordination:

- **Risk Adjustment Focus:** Ensured proper documentation of diabetic complications including nephropathy, retinopathy, and neuropathy
- **HEDIS Focus:** Tracked HbA1c testing, eye exams, nephropathy monitoring, and blood pressure control
- **Clinical Focus:** Implemented evidence-based diabetes management protocols

Provider Education Components:

- Diabetes complication recognition and documentation
- HEDIS measure requirements and clinical significance
- Risk adjustment coding accuracy and compliance

- Care coordination and member engagement strategies

18-Month Results:

- **Risk Adjustment:** Diabetes HCC capture improved from 67% to 89%
- **HEDIS Performance:** Comprehensive diabetes care scores increased from 72% to 85%
- **Clinical Outcomes:** Average HbA1c decreased from 8.2% to 7.6%
- **Financial Impact:** Combined risk adjustment and quality bonus improvements worth $4.8 million annually

Star Ratings Methodology and Risk Adjustment Interactions

The Medicare Advantage Star Ratings program creates quality bonus payments that can significantly exceed risk adjustment impacts, making Star Ratings performance essential for financial sustainability. Understanding how risk adjustment practices affect Star Ratings outcomes enables strategic coordination that maximizes both payment streams.

Quality Bonus Payments range from 0% to 5% of benchmark payments based on overall Star Ratings performance, creating substantial financial incentives for quality improvement. Organizations with 4+ star ratings receive significant bonus payments, while those below 3 stars face potential enrollment restrictions.

Case Mix Adjustment in certain Star Ratings measures accounts for member health status, creating potential interactions with risk adjustment coding practices. More accurate diagnosis coding can improve case mix adjustment while supporting quality measure performance.

Outcome versus Process Measures create different relationships with risk adjustment coding. Process measures depend on accurate member identification through diagnosis codes, while outcome measures may be affected by coding practices that influence case mix adjustment.

Cut Point Methodology determines Star Ratings thresholds based on national performance distributions, meaning that individual organization improvements must exceed industry-wide gains to achieve rating increases. This dynamic creates competitive pressure for continuous improvement.

Display Measures provide early indicators of future rating measures, enabling proactive preparation and improvement initiatives. Organizations must balance current rating optimization with preparation for future measure requirements.

Quality Bonus Payment Optimization

Quality bonus payments create substantial financial opportunities that require strategic planning and coordinated execution across clinical, quality, and risk adjustment domains. Understanding payment mechanics and optimization strategies becomes essential for maximizing these opportunities.

Threshold Achievement requires understanding specific measure thresholds and the relative improvement needed to achieve higher star levels. Organizations must prioritize measures with greatest improvement potential and highest impact on overall ratings.

Measure Weighting affects the relative importance of different quality measures in overall Star Ratings calculations. High-weight measures deserve proportionally greater attention and resource allocation for improvement initiatives.

Cut Point Analysis helps organizations understand their competitive position and the improvements needed to achieve higher ratings. This analysis guides resource allocation and improvement target setting.

Improvement Trajectory Planning requires multi-year strategies that account for measure methodology changes, threshold evolution, and competitive dynamics. Organizations must balance short-term improvements with sustainable long-term performance.

Resource Allocation must consider both the cost of improvement initiatives and the financial benefits of higher Star Ratings. Organizations should prioritize improvements with favorable return on investment profiles.

Case Example 2: Star Ratings Improvement Strategy

Mountain Regional Health Plan faced declining Star Ratings threatening their quality bonus payments:

Baseline Performance (2022):

- Overall Star Rating: 3.0 stars

- Quality bonus payment: 0%
- Risk adjustment coding accuracy: 78%
- Key measure performance gaps in diabetes care, blood pressure control, and medication adherence

Integrated Improvement Strategy:

1. **Member Stratification:** Used risk adjustment data to identify high-risk members for intensive intervention
2. **Provider Engagement:** Aligned risk adjustment and quality improvement incentives in provider contracts
3. **Care Management:** Targeted high-HCC members with quality measure gaps for enhanced services
4. **Technology Enhancement:** Integrated quality measure tracking with risk adjustment coding systems

Coordinated Interventions:

- **Diabetes Management:** Enhanced identification through risk adjustment activities, improved care coordination
- **Hypertension Control:** Used cardiovascular HCC identification to target blood pressure management
- **Medication Adherence:** Leveraged condition identification for targeted pharmacy interventions
- **Preventive Care:** Integrated screening activities with annual wellness visits and HCC capture

Performance Improvement (2024):

- Overall Star Rating: 4.0 stars
- Quality bonus payment: 5% of benchmark ($8.2 million annually)
- Risk adjustment coding accuracy: 92%
- Achieved 4+ stars in 12 of 16 quality measures

Success Factors:

- Leadership commitment to integrated quality and risk adjustment strategy
- Provider incentive alignment across multiple performance domains
- Technology integration enabling coordinated member management

- Cultural transformation emphasizing both quality and accuracy

Care Gap Identification and Member Engagement

Quality measures create systematic approaches for identifying care gaps that often reveal additional risk adjustment opportunities. These care gaps represent missed opportunities for both improved clinical outcomes and enhanced payment accuracy.

Preventive Care Gaps identified through HEDIS measures often reveal members with undiagnosed conditions or inadequate documentation of existing conditions. Mammography screening might identify breast cancer, while colorectal screening could reveal gastrointestinal conditions.

Chronic Disease Management Gaps indicate members whose conditions may be inadequately documented or managed. Missing diabetes eye exams might suggest undocumented diabetic retinopathy, while absent kidney function monitoring could indicate unrecognized nephropathy.

Medication Management Gaps reveal potential diagnosis discrepancies where members receive medications without corresponding diagnosis codes. Antihypertensive medications without hypertension diagnoses suggest documentation opportunities.

Utilization Pattern Analysis identifies members whose healthcare consumption suggests undiagnosed or undocumented conditions. High emergency department utilization might indicate poorly controlled chronic conditions requiring enhanced documentation.

Specialist Referral Patterns provide insights into potential diagnoses not captured in primary care coding. Cardiology referrals might indicate cardiovascular conditions, while endocrinology visits could suggest diabetes complications.

Performance Improvement Strategies and Implementation

Successful quality measure improvement requires systematic approaches that address both clinical care delivery and documentation accuracy. These strategies must integrate quality improvement methodologies with risk adjustment optimization to achieve maximum organizational benefit.

Plan-Do-Study-Act Cycles provide structured approaches for testing improvement interventions and scaling successful strategies. These cycles enable rapid learning and adaptation while minimizing risk of large-scale implementation failures.

Lean Six Sigma methodologies can identify and eliminate waste in quality improvement processes while optimizing workflows for both quality and risk adjustment activities. These approaches emphasize data-driven decision making and process standardization.

Clinical Microsystems approaches focus improvement efforts on small, defined patient populations and provider teams. These approaches enable intensive intervention and measurement while building capabilities for broader implementation.

Change Management processes help organizations adapt clinical practices and documentation behaviors to support both quality improvement and risk adjustment optimization. These processes must address both technical and cultural barriers to change.

Provider Engagement strategies align individual and organizational incentives to support coordinated improvement across quality and risk adjustment domains. These strategies must balance multiple competing priorities while maintaining clinical focus.

Case Example 3: Integrated Care Gap Management

Regional Health Network implemented comprehensive care gap management addressing both quality measures and risk adjustment:

Program Architecture:

1. **Data Integration:** Combined HEDIS measure tracking with risk adjustment analytics
2. **Member Stratification:** Identified high-risk members with multiple care gaps
3. **Provider Workflow:** Integrated gap identification into clinical encounters
4. **Care Management:** Deployed specialized staff for complex member management

Technology Platform:

- Real-time care gap identification during clinical encounters
- Provider dashboards showing both quality measures and risk adjustment opportunities
- Member outreach capabilities for gap closure scheduling

- Automated reporting for performance tracking and improvement

Clinical Integration:

- Enhanced annual wellness visit protocols addressing multiple care gaps
- Specialist referral coordination ensuring appropriate documentation
- Pharmacy collaboration for medication adherence and diagnosis alignment
- Care transition management preventing information loss

24-Month Results:

- **Quality Performance:** HEDIS scores improved average 18 percentage points
- **Risk Adjustment:** HCC capture increased 23% with improved documentation quality
- **Clinical Outcomes:** Reduced emergency department utilization 15%, improved chronic disease control
- **Financial Impact:** Combined quality bonus and risk adjustment improvements worth $14.6 million annually

Sustainability Factors:

- Provider incentive programs supporting both quality and risk adjustment performance
- Technology infrastructure enabling ongoing gap identification and management
- Organizational culture emphasizing integrated performance improvement
- Leadership commitment to resource allocation supporting coordinated strategies

Regulatory Compliance in Integrated Programs

Quality measure and risk adjustment integration must maintain compliance with multiple regulatory frameworks while optimizing performance across both domains. Understanding these compliance requirements prevents program design flaws that could compromise legal or regulatory standing.

Quality Measure Integrity requirements prohibit gaming or manipulation of quality measures through inappropriate clinical or coding practices. Organizations must ensure that risk adjustment activities support rather than compromise quality measure validity.

Clinical Appropriateness standards require that all interventions and documentation improvements reflect genuine clinical needs rather than administrative convenience. These standards prevent inappropriate medicalization of care for administrative purposes.

Member Privacy protections must be maintained across integrated programs that combine clinical, quality, and administrative data. These protections require careful data governance and access control procedures.

Provider Relationships must comply with anti-kickback and inducement regulations when aligning incentives across quality and risk adjustment performance. These relationships require careful legal structuring to ensure compliance.

Documentation Integrity standards apply equally to quality measure and risk adjustment documentation, requiring that all clinical records accurately reflect patient care and clinical decision-making.

Achieving Strategic Excellence Through Integration

Quality measures and Star Ratings create both opportunities and challenges for risk adjustment programs that must balance payment optimization with clinical quality and regulatory compliance. Organizations that successfully integrate these domains achieve sustainable competitive advantages while improving patient outcomes and financial performance.

The evolution toward value-based care increases the importance of quality-risk adjustment integration as organizations must succeed across multiple performance dimensions simultaneously. Professionals who understand these relationships and can develop coordinated strategies will find themselves essential to organizational success.

Integration excellence requires more than technical knowledge—it demands strategic thinking, change management capabilities, and deep understanding of clinical care processes. The professionals who master these competencies will shape the future of healthcare delivery and payment optimization.

Quality Integration Excellence Framework:

- HEDIS measure integration creates direct connections between risk adjustment coding and quality performance requiring coordinated improvement strategies

- Star Ratings methodology and quality bonus payments can significantly exceed risk adjustment impacts making quality performance essential for financial sustainability
- Care gap identification through quality measures reveals additional risk adjustment opportunities while improving clinical outcomes
- Performance improvement strategies must integrate quality improvement methodologies with risk adjustment optimization for maximum organizational benefit
- Regulatory compliance across multiple frameworks requires careful program design that maintains integrity while optimizing performance

Chapter 12: Professional Development and Career Excellence

The risk adjustment field offers exceptional career opportunities for motivated professionals who understand the evolving healthcare landscape and position themselves strategically for advancement. Career success requires more than technical competency—it demands continuous learning, strategic networking, leadership development, and entrepreneurial thinking that creates value for organizations while advancing personal objectives. The intersection of clinical knowledge, business acumen, and regulatory expertise creates unique professional advantages for those who master these domains.

Risk adjustment careers offer multiple progression pathways including individual contributor tracks, management roles, consulting opportunities, and entrepreneurial ventures. Understanding these pathways and developing appropriate capabilities enables strategic career planning that aligns personal interests with market opportunities while building sustainable competitive advantages.

Career Ladder Progression and Strategic Planning

Risk adjustment career progression follows predictable patterns that reward both technical expertise and leadership development. Understanding these patterns enables strategic planning and capability development that accelerates advancement while building valuable expertise.

Entry-Level Positions typically include Risk Adjustment Coder, Clinical Documentation Specialist, and Quality Assurance Analyst roles with salaries ranging from $45,000 to $55,000 annually. These positions require foundational coding knowledge, attention to detail, and willingness to learn complex regulatory requirements.

Mid-Level Roles include Senior Risk Adjustment Specialist, Coding Supervisor, and Analytics Coordinator positions with salaries from $60,000 to $85,000. These roles require advanced technical skills, some supervisory responsibility, and ability to manage complex projects and stakeholder relationships.

Management Positions encompass Risk Adjustment Manager, Director of Clinical Documentation, and Compliance Manager roles with salaries from $90,000 to $130,000. These positions require leadership skills, strategic thinking, and ability to manage teams while achieving organizational objectives.

Senior Leadership Roles include VP of Risk Adjustment, Chief Compliance Officer, and Consulting Principal positions with salaries exceeding $150,000. These roles require executive presence, strategic vision, and ability to influence organizational direction while managing complex stakeholder relationships.

Career advancement typically requires 2-3 years per level with acceleration possible through exceptional performance, additional education, and strategic role selection. Geographic location significantly affects compensation levels, with major metropolitan areas typically offering 15-30% salary premiums.

Case Example 1: Strategic Career Development

Jennifer Martinez progressed from entry-level coder to director-level position within 6 years through strategic planning and capability development:

Career Timeline:

- **Year 1-2:** Risk Adjustment Coder ($48,000) - Mastered CRC certification, achieved 98% coding accuracy
- **Year 3-4:** Senior Specialist ($72,000) - Led documentation improvement project, completed compliance training
- **Year 5-6:** Risk Adjustment Manager ($95,000) - Managed 12-person team, implemented technology solutions
- **Year 7:** Director of Risk Adjustment ($125,000) - Oversees organization-wide risk adjustment operations

Strategic Development Approach:

1. **Technical Excellence:** Maintained top performance in coding accuracy and productivity
2. **Additional Certifications:** Earned CCS, CCA, and compliance certifications
3. **Leadership Experience:** Volunteered for project leadership and team coordination roles
4. **Education Investment:** Completed MBA with healthcare management concentration
5. **Industry Engagement:** Active participation in professional organizations and conferences

Key Success Factors:

- Proactive skill development anticipating future role requirements
- Strong performance in current roles building reputation and opportunities
- Strategic networking creating visibility and advancement possibilities
- Continuous learning maintaining current knowledge and expertise

Salary Negotiation Strategies and Market Understanding

Risk adjustment professionals must understand compensation dynamics and negotiation strategies to maximize earning potential while building sustainable career advancement. Market knowledge, performance documentation, and strategic timing enable effective salary negotiations that reflect professional value.

Market Research should include salary surveys, job posting analysis, and networking insights that provide realistic compensation benchmarks. Geographic differences, organization size, and industry sector significantly affect compensation levels requiring targeted research.

Performance Documentation should quantify professional contributions including cost savings, revenue generation, compliance improvements, and process enhancements. Specific metrics and financial impacts provide compelling negotiation leverage.

Timing Considerations affect negotiation success with annual review cycles, organizational budget processes, and performance achievement creating optimal negotiation windows. Understanding organizational rhythms enables strategic timing.

Value Proposition Development requires articulating unique capabilities, achievements, and future potential in terms that resonate with organizational priorities. Effective propositions connect individual contributions to organizational success.

Negotiation Strategy should consider total compensation including salary, benefits, professional development, and advancement opportunities. Comprehensive approaches often achieve better outcomes than salary-focused negotiations alone.

Alternative Advancement options including lateral moves, expanded responsibilities, and special projects can provide advancement when traditional promotion opportunities are limited.

Advanced Certification Pathways and Professional Credentials

Risk adjustment career advancement benefits significantly from strategic certification acquisition that demonstrates expertise while building marketable credentials. Understanding certification options and their career impacts enables strategic investment in professional development.

Certified Risk Adjustment Coder (CRC) represents the industry standard credential requiring examination passage and ongoing continuing education. This certification provides foundation credibility and entry-level qualification for most risk adjustment positions.

Certified Coding Specialist (CCS) and other AHIMA credentials demonstrate broader coding expertise beyond risk adjustment applications. These certifications provide career flexibility and advancement opportunities in diverse healthcare coding roles.

Certified Compliance and Ethics Professional (CCEP) addresses the growing importance of compliance expertise in risk adjustment operations. This certification becomes increasingly valuable as regulatory enforcement intensifies.

Healthcare Financial Management certifications including CHFP and FHFMA demonstrate business acumen valuable for senior leadership roles. These credentials bridge clinical and financial domains essential for executive advancement.

Project Management certifications including PMP and CAPM provide leadership credentials applicable across healthcare domains. These certifications become essential for management roles involving complex implementation projects.

Data Analytics certifications in areas such as SAS, R, or Python demonstrate analytical capabilities increasingly important in modern risk adjustment operations. These technical skills command premium compensation and advancement opportunities.

Case Example 2: Certification Strategy for Career Advancement

Michael Chen leveraged strategic certification acquisition to transition from coding to analytics leadership:

Certification Timeline:

- **Year 1:** CRC certification establishing risk adjustment foundation
- **Year 2:** CCS certification demonstrating broader coding expertise
- **Year 3:** SAS certification building analytical capabilities

- **Year 4:** CCEP certification addressing compliance requirements
- **Year 5:** MBA completion providing business leadership credentials

Career Progression Results:

- Initial Role: Risk Adjustment Coder ($52,000)
- Mid-Career: Senior Analytics Specialist ($89,000)
- Current Role: Director of Risk Adjustment Analytics ($135,000)

Certification ROI Analysis:

- Total certification investment: $18,500 over 5 years
- Salary increase attributable to certifications: $35,000 annually
- Career advancement acceleration: 2-3 years ahead of typical progression
- Return on investment: 189% annually

Strategic Lessons:

- Certifications should align with career goals and market demands
- Multiple certifications create unique value propositions
- Analytics capabilities command significant premium compensation
- Compliance expertise becomes increasingly valuable

Consulting Opportunities and Independent Practice

Risk adjustment consulting offers exceptional opportunities for experienced professionals who understand market needs and can deliver specialized expertise to organizations facing complex challenges. Consulting success requires business development skills, technical expertise, and ability to manage independent practice operations.

Market Demand for risk adjustment consulting continues to grow as organizations struggle with RADV audits, compliance requirements, and performance optimization. Specialized expertise in areas such as audit preparation, technology implementation, and provider education commands premium rates.

Service Offerings can include compliance assessments, audit preparation, provider training, technology selection, and performance improvement. Successful consultants develop signature methodologies and intellectual property that differentiate their services.

Rate Structures typically range from $150-$300 per hour for project work, with retainer arrangements and performance-based compensation creating additional opportunities. Experienced consultants with specialized expertise can command rates exceeding $400 per hour.

Business Development requires networking, content marketing, and relationship building that establishes credibility and generates referrals. Speaking engagements, industry publications, and professional organization participation create visibility and opportunity.

Practice Management involves client relationship management, project delivery, financial administration, and business planning. Successful independent consultants develop systematic approaches to practice operations.

Partnership Opportunities with larger consulting firms, technology vendors, and healthcare organizations can provide market access and capability enhancement. These partnerships often offer paths to larger engagements and specialized expertise development.

Industry Networking and Professional Relationships

Professional networking creates career opportunities, knowledge sharing, and business development that accelerate advancement while building industry expertise. Strategic networking requires intentional relationship building across multiple professional domains.

Professional Organizations including AAPC, AHIMA, and specialized risk adjustment groups provide networking platforms, educational opportunities, and industry leadership positions. Active participation creates visibility and relationship building opportunities.

Conference Participation through attendance, speaking, and exhibiting creates industry exposure while building expertise and relationships. Strategic conference selection and preparation maximize networking return on investment.

Online Networking through LinkedIn, professional forums, and industry publications enables relationship building and thought leadership development. Consistent engagement and value creation build professional reputation and opportunities.

Mentorship Programs both as mentee and mentor create learning opportunities while building professional relationships. These programs provide career guidance and industry insights valuable for advancement.

Industry Collaboration through committees, working groups, and standards development creates deep industry engagement while building expertise and relationships. These activities often lead to speaking opportunities and thought leadership recognition.

Case Example 3: Network-Driven Career Transformation

Susan Park leveraged strategic networking to transition from health plan employee to successful independent consultant:

Networking Strategy:

1. **Professional Organization Leadership:** Served on AAPC Risk Adjustment Committee
2. **Conference Speaking:** Developed expertise presentations on RADV audit preparation
3. **Content Creation:** Published articles on risk adjustment best practices
4. **Mentorship Program:** Mentored emerging professionals while learning from industry veterans
5. **Strategic Partnerships:** Developed relationships with technology vendors and consulting firms

Business Development Results:

- Industry recognition as thought leader in RADV compliance
- Speaking engagements at 12 major conferences annually
- Consulting client base including 15 health plans and 25 provider organizations
- Annual consulting revenue exceeding $450,000

Network-Generated Opportunities:

- Board appointment to risk adjustment standards organization
- Technology vendor advisory roles providing additional income
- Teaching opportunities at professional education programs
- Book publishing contract for risk adjustment compliance guide

Relationship Management Approach:

- Consistent value delivery through knowledge sharing and expertise
- Strategic follow-up maintaining relationships beyond immediate opportunities

- Collaborative approach building win-win professional relationships
- Reputation management through ethical practice and professional excellence

Building Industry Expertise and Thought Leadership

Thought leadership creates professional differentiation, advancement opportunities, and business development while contributing to industry knowledge and standards development. Building expertise requires strategic focus, consistent content creation, and authentic value delivery.

Specialization Development in areas such as RADV audits, specific clinical conditions, or technology implementation creates differentiated expertise that commands premium opportunities. Deep specialization often proves more valuable than broad generalist knowledge.

Content Creation through articles, presentations, webinars, and social media builds professional reputation while demonstrating expertise. Consistent content creation establishes thought leadership and industry recognition.

Research and Analysis of industry trends, regulatory changes, and best practices creates valuable insights that benefit both personal advancement and industry development. Original research often generates speaking and publication opportunities.

Teaching and Education through professional organizations, academic institutions, and corporate training programs builds expertise while creating industry impact. Educational roles often lead to additional opportunities and professional recognition.

Industry Standards participation through committee work, guideline development, and standards creation contributes to industry advancement while building expertise and relationships. These activities create significant professional credibility.

Innovation and Development of new methodologies, technologies, and approaches creates intellectual property while advancing industry capabilities. Innovation often leads to consulting opportunities and professional recognition.

Creating Lasting Professional Success

Risk adjustment careers offer exceptional opportunities for professionals who understand the evolving healthcare landscape and position themselves strategically for advancement.

Success requires combination of technical expertise, leadership capabilities, business acumen, and strategic networking that creates value for organizations while advancing personal objectives.

The field continues to evolve rapidly as regulatory requirements change, technology advances, and healthcare delivery models transform. Professionals who maintain current knowledge while developing complementary capabilities will find themselves essential to organizational success and well-positioned for leadership roles.

Career excellence in risk adjustment requires more than individual achievement—it demands contribution to industry advancement, ethical practice, and knowledge sharing that benefits the broader professional community. The professionals who embrace these responsibilities while achieving personal success will shape the future of risk adjustment and healthcare delivery.

Professional Excellence Framework:

- Career ladder progression rewards both technical expertise and leadership development with predictable advancement patterns and compensation growth
- Salary negotiation success requires market knowledge, performance documentation, and strategic value proposition development
- Advanced certification pathways provide strategic credentialing that demonstrates expertise while building marketable credentials
- Consulting opportunities offer exceptional advancement potential for experienced professionals with specialized expertise and business development capabilities
- Industry networking and thought leadership create career opportunities while contributing to professional community advancement

Chapter 13: Strategic Exam Preparation

The Certified Risk Adjustment Coder examination represents more than a credentialing milestone—it serves as the gateway to career advancement, professional recognition, and financial opportunity in the rapidly growing risk adjustment field. Successful examination performance requires systematic preparation that combines content mastery, strategic study approaches, and test-taking optimization. The investment in thorough preparation pays substantial returns through certification achievement and the career opportunities that follow.

Modern exam preparation benefits from evidence-based learning methodologies that maximize retention while optimizing study efficiency. Understanding cognitive science principles, implementing spaced repetition systems, and utilizing active recall techniques can dramatically improve preparation outcomes while reducing total study time. These approaches transform preparation from passive content review to active knowledge construction that supports both examination success and long-term professional competency.

12-Week Study Schedule Development and Optimization

Effective CRC exam preparation requires structured scheduling that balances comprehensive content coverage with adequate review and practice opportunities. A 12-week preparation timeline provides sufficient depth while maintaining motivation and focus throughout the study period.

Weeks 1-3: Foundation Building The initial phase establishes fundamental knowledge across all exam domains while identifying strength and weakness areas. This phase emphasizes understanding rather than memorization, creating conceptual frameworks that support advanced learning.

Week 1 focuses on risk adjustment fundamentals including Medicare Advantage basics, payment methodologies, and regulatory frameworks. Students should master basic terminology, understand key stakeholders, and grasp the economic foundations of risk adjustment.

Week 2 addresses risk adjustment models including CMS-HCC, HHS-HCC, and specialized models. Students should understand model architecture, coefficient systems, RAF score calculations, and hierarchical relationships.

Week 3 covers ICD-10-CM foundations including coding guidelines, anatomy and physiology basics, and medical terminology. Students should achieve comfort with code structure, guideline application, and basic pathophysiology concepts.

Weeks 4-8: Core Content Mastery The middle phase provides deep coverage of clinical content areas that comprise the majority of examination questions. This phase emphasizes application of knowledge to realistic scenarios and complex coding situations.

Week 4 addresses cardiovascular conditions including heart failure classification, coronary disease coding, and arrhythmia documentation. Students should master the distinctions between acute and chronic conditions, systolic versus diastolic heart failure, and complication coding.

Week 5 covers diabetes and endocrine conditions including complication recognition, hierarchy relationships, and documentation requirements. Students should understand the constrained coefficient system and complication categorization.

Week 6 focuses on neurological and mental health conditions including stroke sequelae, dementia coding, and psychiatric condition classification. Students should master the expanded neurological categories and mental health documentation requirements.

Week 7 addresses respiratory, immune, and other system conditions including COPD classification, autoimmune disorders, and cancer coding. Students should understand staging requirements, treatment status implications, and transplant-related coding.

Week 8 covers compliance and ethics including RADV audit requirements, fraud prevention, and ethical decision-making frameworks. Students should understand regulatory requirements and professional standards.

Weeks 9-11: Integration and Application The advanced phase integrates knowledge across domains while emphasizing complex scenarios and practical application. This phase simulates examination conditions while reinforcing key concepts through varied practice formats.

Week 9 provides comprehensive case studies that integrate multiple clinical conditions and coding challenges. Students should practice complex scenario analysis and multi-system coding decisions.

Week 10 emphasizes RADV audit preparation and compliance applications including documentation analysis, audit response strategies, and quality assurance processes.

Week 11 focuses on analytics and technology applications including predictive modeling concepts, quality measure integration, and population health management.

Week 12: Final Review and Exam Preparation The final week consolidates learning through intensive review, final practice examinations, and mental preparation for test day. This phase emphasizes confidence building and test-taking strategy refinement.

Spaced Repetition Implementation and Memory Optimization

Spaced repetition represents one of the most powerful learning techniques available, improving long-term retention by up to 200% compared to traditional study methods. Understanding and implementing spaced repetition principles can dramatically improve examination preparation efficiency while building lasting professional knowledge.

Forgetting Curve Management Hermann Ebbinghaus demonstrated that memory retention follows predictable decay patterns, with most information forgotten within 24 hours without reinforcement. Spaced repetition combats this decay by scheduling review sessions at optimal intervals that strengthen memory consolidation.

Initial learning should be followed by review sessions at 1 day, 3 days, 7 days, 14 days, and 30 days to achieve optimal retention. This schedule leverages the brain's natural memory consolidation processes while preventing the cognitive overload associated with massed practice.

Difficulty-Based Spacing Material difficulty should influence review scheduling, with challenging concepts requiring more frequent review than easily mastered content. Students should adjust review intervals based on recall success, extending intervals for well-learned material while shortening them for challenging concepts.

Active Recall Integration Spaced repetition works best when combined with active recall techniques that require memory retrieval rather than passive recognition. Students should use flashcards, practice questions, and self-testing rather than simply re-reading notes or textbooks.

Technology-Enhanced Implementation Digital spaced repetition systems can automate scheduling and track performance across thousands of individual facts and concepts. These systems adapt to individual learning patterns while ensuring optimal review timing for maximum retention.

Case Example 1: Spaced Repetition Study Program

David Thompson implemented a systematic spaced repetition program for CRC exam preparation:

Content Organization:

- 2,847 individual flashcards covering all exam domains
- Hierarchical organization by body system and complexity level
- Integration with practice questions and case studies
- Performance tracking and adaptive scheduling

Implementation Schedule:

- Daily review sessions averaging 45 minutes
- Weekend intensive sessions for new content introduction
- Weekly performance assessment and schedule adjustment
- Monthly comprehensive review and weak area identification

Review Methodology:

- Active recall through self-testing before answer review
- Immediate feedback and explanation for incorrect responses
- Difficulty rating adjustment based on recall success
- Integration with practice examinations and case studies

Results After 10 Weeks:

- Overall retention rate: 94% (vs. 67% baseline)
- Practice exam scores: 89% average (vs. 72% initial)
- Time efficiency: 40% reduction in total study time
- Confidence level: Significantly improved across all domains

Success Factors:

- Consistent daily implementation without exception

- Technology support for scheduling and tracking
- Integration with other study methodologies
- Regular assessment and program adjustment

Active Recall Techniques and Application Strategies

Active recall techniques require learners to retrieve information from memory rather than simply recognizing correct answers, creating stronger neural pathways and improving long-term retention. Understanding and implementing active recall can improve examination performance while building practical professional competency.

Self-Testing Protocols Students should regularly test themselves using flashcards, practice questions, and scenario analysis without referring to reference materials. This approach identifies knowledge gaps while strengthening memory retrieval pathways essential for examination success.

Explanation Generation Students should practice explaining complex concepts in their own words, as if teaching someone else. This technique reveals understanding gaps while building the conceptual frameworks necessary for application to novel situations.

Case Study Analysis Students should work through realistic clinical scenarios that require integration of knowledge across multiple domains. This approach simulates examination conditions while building practical problem-solving capabilities.

Peer Teaching Students should seek opportunities to explain concepts to study partners or colleagues, as teaching requires deep understanding and reveals knowledge gaps not apparent through individual study.

Question Generation Students should create their own practice questions based on study material, as question development requires understanding of both content and application principles essential for examination success.

Time Management Strategies and Study Efficiency

Effective time management maximizes study productivity while maintaining work-life balance throughout the preparation period. Understanding productivity principles and implementing systematic approaches can improve study outcomes while reducing stress and burnout.

Pomodoro Technique Implementation The Pomodoro Technique uses 25-minute focused study sessions followed by 5-minute breaks to maintain concentration while preventing mental fatigue. This approach can improve study efficiency by up to 40% compared to extended study sessions.

Energy Management Students should schedule challenging material during peak energy periods while reserving review activities for lower-energy times. Understanding personal circadian rhythms enables optimization of study scheduling for maximum effectiveness.

Environment Optimization Study environments should minimize distractions while providing necessary resources and comfort. Consistent study locations create environmental cues that facilitate focus and learning state activation.

Goal Setting and Progress Tracking Students should establish specific, measurable study goals with regular progress assessment. Clear objectives provide motivation while enabling course correction when progress falls behind expectations.

Stress Management Students should implement stress reduction techniques including exercise, adequate sleep, and relaxation practices that support cognitive performance while maintaining overall well-being.

Test-Taking Strategies and Performance Optimization

Examination performance depends not only on content knowledge but also on strategic test-taking approaches that maximize score potential while managing time effectively. Understanding examination format and implementing proven strategies can improve performance beyond content mastery alone.

Question Analysis Techniques Students should read questions carefully, identifying key words and concepts before reviewing answer choices. This approach prevents premature answer selection while ensuring complete question comprehension.

Elimination Strategies Students should systematically eliminate obviously incorrect answers before selecting from remaining choices. This approach improves odds even when complete certainty is unavailable.

Time Allocation Management Students should allocate time proportionally across examination sections while maintaining awareness of overall progress. This approach prevents time shortages while ensuring adequate attention to all content areas.

Confidence Level Assessment Students should mark questions where confidence is low for potential review if time permits. This approach enables strategic time allocation while building confidence through question completion.

Stress and Anxiety Management Students should implement breathing techniques, positive self-talk, and relaxation strategies that maintain calm focus throughout the examination period.

Case Example 2: Comprehensive Test-Taking Strategy

Maria Rodriguez developed and implemented systematic test-taking strategies that improved her practice exam performance:

Pre-Exam Preparation:

- Examination format familiarization through official practice tests
- Time allocation planning based on question difficulty distribution
- Stress management technique development and practice
- Physical preparation including nutrition and sleep optimization

During-Exam Strategy:

- Question analysis protocol: Read completely, identify key concepts, eliminate wrong answers
- Time management: 90 seconds average per question with difficult question flagging
- Confidence tracking: Mark uncertain answers for potential review
- Stress management: Deep breathing between sections, positive self-talk

Review Process:

- Systematic review of flagged questions when time permits
- Answer change protocol: Only change when confident of error
- Final review: Check for incomplete answers and obvious errors

Performance Results:

- Practice exam improvement: From 78% to 91% average
- Time management: Completed all questions with 15 minutes remaining
- Confidence level: Reduced test anxiety and improved focus

- Actual exam performance: Passed with 89% score

Strategy Refinement:

- Weekly strategy assessment and adjustment during preparation
- Practice test analysis identifying strategy effectiveness
- Peer consultation regarding successful approaches
- Continuous improvement based on performance feedback

Comprehensive Review and Final Preparation

The final preparation phase consolidates learning while building confidence and optimizing performance for examination day. This phase requires systematic review that reinforces key concepts while addressing remaining knowledge gaps.

Content Consolidation Students should create comprehensive summaries that integrate knowledge across all examination domains while highlighting key relationships and applications. These summaries serve as final review materials while reinforcing conceptual frameworks.

Weak Area Remediation Students should identify and address remaining knowledge gaps through targeted study and practice. This remediation should focus on high-value content areas with significant examination representation.

Practice Examination Completion Students should complete multiple full-length practice examinations under realistic time constraints to build endurance while identifying final preparation needs. These examinations should simulate actual testing conditions as closely as possible.

Mental Preparation Students should develop positive mental frameworks that build confidence while managing anxiety. Visualization techniques, positive affirmations, and success planning create mental readiness for examination performance.

Physical Preparation Students should optimize physical condition through adequate sleep, proper nutrition, and stress management that support cognitive performance during examination day.

Achieving Certification Excellence Through Strategic Preparation

CRC examination success requires more than content knowledge—it demands strategic preparation that optimizes learning while building test-taking capabilities essential for peak performance. The investment in systematic preparation pays substantial returns through certification achievement and the career opportunities that follow.

Understanding cognitive science principles and implementing evidence-based study techniques can dramatically improve preparation efficiency while building lasting professional competency. The professionals who master these approaches achieve certification success while developing learning capabilities that support lifelong career advancement.

Examination preparation represents an investment in professional future that requires commitment, discipline, and strategic thinking. The professionals who approach preparation systematically while maintaining focus on long-term career objectives will find themselves well-positioned for success in the dynamic risk adjustment field.

Strategic Preparation Excellence Framework:

- 12-week study schedules provide structured progression through foundation building, content mastery, integration, and final review phases
- Spaced repetition implementation dramatically improves retention while optimizing study efficiency through evidence-based scheduling
- Active recall techniques build stronger memory pathways while developing practical application capabilities essential for professional success
- Time management strategies maximize study productivity while maintaining work-life balance throughout the preparation period
- Test-taking optimization improves performance beyond content mastery through strategic question analysis and examination management

Chapter 14: Final Review and Practice Exams

The culmination of CRC exam preparation requires intensive consolidation that transforms accumulated knowledge into examination-ready competency while building confidence for peak performance. This final phase serves dual purposes: identifying and addressing remaining knowledge gaps while optimizing test-taking capabilities through realistic practice experiences. Success in this critical phase determines whether months of preparation translate into certification achievement and career advancement.

Effective final review goes beyond simple content repetition to include strategic knowledge organization, weakness remediation, and performance optimization that ensures examination readiness. Understanding how to maximize this crucial period enables candidates to approach their examination with confidence while demonstrating mastery across all required competency domains.

Domain-Weighted Review and Knowledge Consolidation

The CRC examination distributes questions across defined content domains that require proportional attention during final review. Understanding these weightings enables strategic time allocation that maximizes preparation effectiveness while ensuring adequate coverage of high-impact areas.

Risk Adjustment Methodology (25% of examination) This domain encompasses payment models, RAF score calculations, hierarchical relationships, and regulatory frameworks. Final review should emphasize model comparisons, coefficient applications, and real-world scenarios that demonstrate understanding of payment mechanics.

Students should master CMS-HCC Model V28 specifics including all 115 payment categories, hierarchy relationships, and constrained coefficient applications. Understanding demographic factors, interaction terms, and blended payment calculations becomes essential for examination success.

ICD-10-CM Coding (25% of examination) This domain covers diagnosis coding specific to risk adjustment applications including guidelines application, specificity requirements, and clinical documentation correlation. Final review should emphasize high-value diagnosis categories and common coding scenarios.

Students should master the approximately 7,770 diagnosis codes that map to HCC categories while understanding coding guidelines application and documentation requirements. Emphasis should be placed on complex coding scenarios and hierarchy relationships.

Documentation & Compliance (20% of examination) This domain addresses MEAT criteria, RADV audit requirements, provider qualifications, and compliance standards. Final review should emphasize practical application of documentation standards and audit preparation strategies.

Students should master MEAT criteria application across diverse clinical scenarios while understanding acceptable provider types and encounter requirements. RADV audit preparation and compliance standards require particular attention.

Medical Terminology (15% of examination) This domain covers anatomy, physiology, pathophysiology, and medical terminology relevant to risk adjustment coding. Final review should emphasize clinical knowledge application and disease process understanding.

Students should master organ system relationships, disease progression patterns, and clinical terminology that supports accurate diagnosis recognition and coding assignment.

HCC Categories & Hierarchies (15% of examination) This domain addresses specific HCC categories, hierarchical relationships, and clinical applications. Final review should emphasize category-specific requirements and hierarchy navigation.

Students should master all HCC categories with particular attention to high-value conditions and complex hierarchical relationships that affect payment outcomes.

Case Example 1: Domain-Weighted Study Plan

Angela Foster implemented a domain-weighted final review strategy based on examination content distribution:

Week 1 Review Allocation:

- **Risk Adjustment Methodology (25%):** 10 hours focused on model comparisons, RAF calculations, and payment mechanics
- **ICD-10-CM Coding (25%):** 10 hours emphasizing high-value codes and complex scenarios

- **Documentation & Compliance (20%):** 8 hours covering MEAT criteria and RADV requirements
- **Medical Terminology (15%):** 6 hours reviewing anatomy and pathophysiology
- **HCC Categories & Hierarchies (15%):** 6 hours mastering category relationships

Study Materials by Domain:

- **Methodology:** Official CMS documentation, payment model specifications, calculator practice
- **Coding:** ICD-10-CM manual, coding scenarios, guideline applications
- **Documentation:** RADV audit examples, MEAT criteria practice, compliance cases
- **Terminology:** Medical terminology flashcards, anatomy reviews, pathophysiology summaries
- **HCC Categories:** Category-specific fact sheets, hierarchy charts, interaction examples

Performance Tracking:

- Daily domain-specific practice questions with performance analysis
- Weekly domain-weighted practice examinations
- Weak area identification and targeted remediation
- Progress monitoring against examination benchmarks

Results After 3 Weeks:

- Overall practice exam scores improved from 82% to 94%
- Domain-specific performance: All domains above 90%
- Confidence level significantly increased across all areas
- Time management optimized for examination conditions

High-Yield Fact Summaries and Quick Reference

Final review efficiency benefits from concentrated fact summaries that highlight essential information requiring immediate recall during examination situations. These summaries should emphasize high-impact concepts, common testing areas, and frequently missed details.

HCC Category Quick Reference Students should memorize high-value HCC categories with their coefficients, common diagnoses, and hierarchy relationships. This information enables rapid question analysis and answer selection during time-pressured examination conditions.

Critical categories include HCC 221 (Acute MI - 0.323), HCC 225 (CHF - 0.323), HCC 35 (Diabetes with complications - 0.318), HCC 326 (CKD stage 3-4 - 0.237), and HCC 155 (Major depression - 0.309).

MEAT Criteria Applications Students should master specific MEAT criteria applications across diverse clinical scenarios. Understanding what constitutes acceptable monitoring, evaluation, assessment, and treatment becomes essential for documentation questions.

Monitor examples include vital signs, laboratory results, symptom assessment, and functional status evaluation. Evaluate includes diagnostic test interpretation, treatment effectiveness assessment, and condition progression analysis.

Documentation Requirements Students should memorize face-to-face encounter requirements, acceptable provider types, and timing restrictions that affect diagnosis validity. These requirements frequently appear in examination questions requiring immediate recall.

Coding Guidelines Students should master key ICD-10-CM guidelines affecting risk adjustment coding including combination codes, multiple coding requirements, and sequencing rules. These guidelines influence many examination questions.

Case Example 2: Quick Reference Development

Robert Kim created comprehensive quick reference materials for final review:

HCC Category Cards:

- 115 individual cards with category number, coefficient, common diagnoses, and hierarchy relationships
- Color-coded by body system and value level
- Practice scenarios on reverse side for application testing
- Mobile-friendly format for convenient review

MEAT Criteria Examples:

- 50 clinical scenarios with MEAT criteria identification

- Body system organization with specialty-specific examples
- Documentation quality assessment rubrics
- Common failure patterns and correction strategies

Coding Guidelines Summary:

- Essential guideline excerpts organized by application frequency
- Risk adjustment specific interpretations and examples
- Common error patterns and prevention strategies
- Quick lookup format for rapid reference

Usage and Results:

- Daily 15-minute review sessions during commute
- Integration with practice question analysis
- Peer study group sharing and discussion
- Examination day reference for final preparation
- Performance improvement: 15% increase in speed and accuracy

Last-Minute Study Tips and Optimization

The final days before examination require strategic approaches that consolidate learning while maintaining peak mental and physical condition. Understanding how to optimize this crucial period prevents cramming while ensuring readiness for examination performance.

Content Consolidation Students should focus on review and reinforcement rather than new learning during the final 48 hours. Attempting to master new material creates confusion while undermining confidence in previously learned content.

Review should emphasize high-yield facts, common testing areas, and weak spots identified through practice examinations. Students should avoid detailed study of unfamiliar material in favor of reinforcing established knowledge.

Practice Question Strategy Students should continue practice questions at reduced intensity to maintain mental sharpness while building confidence. Questions should focus on realistic examination scenarios rather than obscure edge cases.

Incorrect answers should receive immediate attention with explanation review and concept reinforcement. Students should maintain error logs to prevent repetition of common mistakes.

Mental Preparation Students should implement positive visualization, confidence building exercises, and stress management techniques that optimize mental state for examination performance.

Adequate sleep, proper nutrition, and light physical activity support cognitive function while maintaining overall well-being during this critical period.

Physical Preparation Students should ensure physical readiness through examination logistics planning, transportation arrangements, and arrival scheduling that prevents last-minute stress.

Required materials including identification, calculators, and writing implements should be prepared in advance with backup options available.

Exam Day Strategies and Performance Management

Examination day success requires strategic approaches that optimize performance while managing stress and time effectively. Understanding proven strategies and implementing systematic approaches can improve performance beyond content mastery alone.

Pre-Examination Routine Students should develop consistent pre-examination routines that promote calm focus while ensuring physical and mental readiness. These routines should begin the evening before examination and continue through arrival at the testing center.

Morning preparation should include light nutrition, adequate hydration, and stress management techniques that support cognitive performance without creating nervousness or distraction.

During Examination Strategy Students should implement systematic question analysis approaches that maximize accuracy while managing time effectively. Reading questions completely, identifying key concepts, and eliminating incorrect answers before selection improves performance.

Time management requires balance between thorough question analysis and adequate progress through the examination. Students should monitor timing regularly while avoiding panic about pace.

Stress Management Students should implement breathing techniques, positive self-talk, and relaxation strategies that maintain calm focus throughout the examination period. Stress management becomes particularly important when encountering difficult questions or time pressure.

Answer Selection Protocol Students should develop consistent approaches to answer selection that balance confidence with systematic analysis. Changing answers should only occur when clearly identifying errors in initial selection.

Case Example 3: Examination Day Excellence

Lisa Wang implemented comprehensive examination day strategies that optimized her performance:

Pre-Examination Preparation:

- Evening before: Light review of quick reference materials, early bedtime
- Morning routine: Healthy breakfast, light exercise, arrival 30 minutes early
- Mental preparation: Positive visualization, stress management techniques

During Examination Management:

- Question analysis: Read completely, identify key words, eliminate wrong answers
- Time allocation: 90 seconds per question with difficult question flagging
- Stress management: Deep breathing between sections, positive self-talk
- Progress monitoring: Regular time checks without panic

Performance Results:

- Completed all questions with 12 minutes remaining
- Used remaining time for flagged question review
- Maintained calm focus throughout examination
- Achieved 91% score, exceeding target of 85%

Success Factors:

- Systematic preparation and practice of strategies
- Consistent implementation without deviation
- Confidence from thorough preparation
- Stress management preventing performance degradation

Post-Examination Planning and Career Development

Successful examination completion marks the beginning rather than the end of professional development in risk adjustment. Understanding post-certification opportunities and planning career advancement enables maximum benefit from certification achievement.

Immediate Post-Examination Steps Students should begin job searching and career advancement planning immediately after examination completion. The job market for certified risk adjustment coders remains strong with numerous opportunities for advancement.

Continuing Education Planning Students should plan ongoing professional development that maintains current knowledge while building advanced capabilities. The risk adjustment field evolves rapidly requiring continuous learning and adaptation.

Professional Networking Students should engage with professional organizations and industry networks that provide advancement opportunities while building relationships essential for career growth.

Specialization Development Students should consider areas of specialization that align with personal interests and market opportunities. Specialization in areas such as RADV audits, analytics, or compliance creates competitive advantages.

Long-Term Career Planning Students should develop strategic career plans that leverage certification while building toward leadership roles and advanced opportunities in the evolving risk adjustment field.

Completing the Journey to Professional Excellence

CRC examination success represents a significant milestone in risk adjustment career development, opening doors to advancement opportunities while demonstrating professional competency. The final review phase determines whether months of preparation translate into certification achievement and career advancement.

Effective final preparation requires systematic approaches that consolidate knowledge while optimizing examination performance. Understanding domain weightings, implementing strategic review techniques, and managing examination day performance can make the difference between success and disappointment.

Certification achievement marks the beginning of lifelong learning and professional development in the dynamic risk adjustment field. The professionals who approach certification as a foundation rather than destination will find themselves well-positioned for leadership roles and career advancement in this growing field.

Final Excellence Framework:

- Domain-weighted review ensures proportional attention to examination content areas with strategic time allocation maximizing preparation effectiveness
- High-yield fact summaries provide concentrated information requiring immediate recall during examination situations
- Last-minute optimization focuses on consolidation and confidence building rather than new learning during the final preparation period
- Examination day strategies optimize performance through systematic question analysis and stress management techniques
- Post-certification planning leverages achievement for career advancement while building toward long-term professional objectives

Chapter 15: Practice Questions

Risk Adjustment Methodology

Foundation Level

Question 1 (Multiple Choice) What is the primary purpose of risk adjustment in Medicare Advantage plans?

A) To reduce healthcare costs for all members B) To ensure appropriate payment based on member health status C) To eliminate high-risk members from enrollment D) To standardize treatment protocols across all providers

Answer: B.

Risk adjustment ensures that Medicare Advantage plans receive appropriate payments based on the health status and expected costs of their enrolled members. Plans with sicker populations receive higher payments to account for increased medical costs.

Reference: CMS-HCC Model Documentation V28

Common Error: Confusing risk adjustment with cost containment measures.

Question 2 (Multiple Choice) The Risk Adjustment Factor (RAF) score represents:

A) The percentage of members who will require hospitalization B) A multiplier applied to the base payment rate C) The average age of plan members D) The number of chronic conditions per member

Answer: B.

The RAF score is a multiplier that adjusts the base payment rate up or down based on the predicted costs for a member's demographic and health status profile.

Reference: 42 CFR 422.308

Common Error: Thinking RAF scores represent percentages or counts rather than payment multipliers.

Question 3 (Select All That Apply) Which of the following are components used in calculating RAF scores?

A) Age B) Gender
C) Medicaid eligibility D) Geographic location E) Hierarchical Condition Categories (HCCs)

Answer: A, B, C, E.

RAF scores incorporate demographic factors (age, gender), Medicaid eligibility status, and HCCs. Geographic location affects payment rates but is not part of the RAF calculation itself.

Reference: CMS-HCC Model V28 Technical Documentation

Common Error: Including geographic adjustments as part of RAF rather than separate payment adjustments.

Question 4 (Multiple Choice) How frequently must HCC-eligible diagnoses be documented to continue generating payments?

A) Once per lifetime B) Annually C) Every six months D) Monthly

Answer: B.

HCC diagnoses must be documented at least once per calendar year during a face-to-face encounter with an acceptable provider to continue generating risk adjustment payments.

Reference: CMS Risk Adjustment Data Technical Assistance

Common Error: Assuming once documented, conditions continue to generate payments indefinitely.

Question 5 (Scenario-Based) A 67-year-old female Medicare Advantage member has documented diabetes with chronic kidney disease. Her base demographic RAF is 0.45. The diabetes with CKD adds 0.318 to her RAF score. What is her total RAF score?

A) 0.45 B) 0.318 C) 0.768 D) 0.132

Answer: C.

RAF scores are additive. Base demographic (0.45) + HCC value (0.318) = Total RAF (0.768).

Reference: CMS-HCC Model V28 Coefficient Tables

Common Error: Multiplying RAF components instead of adding them.

Question 6 (Multiple Choice) Which model is used for Medicare Advantage risk adjustment?

A) HHS-HCC B) CMS-HCC C) CDPS D) APR-DRG

Answer: B.

The CMS-HCC (Centers for Medicare & Medicaid Services Hierarchical Condition Categories) model is specifically used for Medicare Advantage risk adjustment.

Reference: 42 CFR 422.308(c)

Common Error: Confusing CMS-HCC with HHS-HCC, which is used for ACA marketplace plans.

Question 7 (Multiple Choice) What does "prospective payment" mean in risk adjustment?

A) Payments are made after services are rendered B) Payments are based on predicted future costs C) Payments are adjusted retrospectively D) Payments are made in advance of enrollment

Answer: B.

Prospective payment means payments are made based on predicted future medical costs using current year demographic and prior year diagnosis information.

Reference: Medicare Managed Care Manual Chapter 7

Common Error: Confusing prospective with advance payment timing.

Question 8 (Multiple Choice) The current CMS-HCC model version is:

A) V24 B) V26 C) V28 D) V30

Answer: C.

CMS-HCC Model V28 is the current version implemented for payment year 2024 and beyond.

Reference: CMS-HCC Model V28 Announcement

Common Error: Using outdated model version information.

Question 9 (Select All That Apply) Which populations use CMS-HCC risk adjustment? (Select all that apply)

A) Medicare Advantage members B) ACA marketplace enrollees C) Medicare Part D standalone prescription drug plans D) Medicaid managed care members E) Employer-sponsored health plans

Answer: A, C.

CMS-HCC is used for Medicare Advantage and Medicare Part D risk adjustment. ACA marketplace uses HHS-HCC, while Medicaid and employer plans may use other models.

Reference: CMS Risk Adjustment Overview

Common Error: Assuming all government health programs use the same risk adjustment model.

Question 10 (Multiple Choice) What is the maximum number of HCCs that can contribute to a member's RAF score?

A) 5 B) 10 C) 85 D) No maximum limit

Answer: D.

There is no maximum limit to the number of HCCs that can contribute to a member's RAF score, though hierarchies prevent double-counting related conditions.

Reference: CMS-HCC Model V28 Technical Documentation

Common Error: Believing there are caps on the number of contributing HCCs.

Question 11 (Calculation Exercise) Calculate the total RAF score for a 72-year-old male with the following:

- Base demographic RAF: 0.583
- Diabetes with complications (HCC 35): 0.318
- COPD (HCC 168): 0.328

A) 0.583 B) 0.646 C) 1.229 D) 0.323

Answer: C.

Total RAF = 0.583 (demographic) + 0.318 (HCC 35) + 0.328 (HCC 168) = 1.229.

Reference: CMS-HCC Model V28 Coefficient Tables

Common Error: Averaging the values instead of adding them.

Question 12 (Multiple Choice) Risk adjustment data is submitted to CMS through which system?

A) RAPS (Risk Adjustment Processing System) B) EDPS (Encounter Data Processing System) C) Both RAPS and EDPS D) HPMS (Health Plan Management System)

Answer: B.

As of 2016, encounter data through EDPS is the primary submission method, with RAPS being phased out for most submissions.

Reference: CMS EDPS User Guide

Common Error: Still referring to RAPS as the current primary submission method.

Question 13 (Multiple Choice) What does "data year" refer to in risk adjustment?

A) The year payment is made B) The year diagnoses are documented C) The year the member enrolled D) The year the audit occurs

Answer: B.

Data year refers to the calendar year in which the diagnoses used for risk adjustment were documented and reported.

Reference: CMS Risk Adjustment Data Validation Guidance

Common Error: Confusing data year with payment year.

Question 14 (Select All That Apply) Which factors can affect a member's demographic RAF score? (Select all that apply)

A) Age B) Gender C) Original reason for Medicare entitlement D) Dual eligibility status E) Geographic location

Answer: A, B, C, D.

Demographic RAF includes age, gender, original reason for Medicare entitlement (disability vs. age), and dual eligibility status. Geographic location affects payment rates separately.

Reference: CMS-HCC Model V28 Demographic Categories

Common Error: Including geographic factors in demographic RAF calculation.

Question 15 (Multiple Choice) The term "suspect" in risk adjustment refers to:

A) Fraudulent diagnosis codes B) Diagnoses that may be missing from current year data C) Incorrect provider credentials D) Duplicate member enrollments

Answer: B.

Suspects are diagnoses from the prior payment year that may be missing from current year data, potentially indicating gaps in documentation or care.

Reference: CMS Suspect Processing Guidelines

Common Error: Assuming suspects indicate fraudulent activity.

Question 16 (Multiple Choice) What is the payment year in risk adjustment?

A) The year diagnoses are documented B) The year payments are calculated and made C) The year members enroll in the plan D) The year services are provided

Answer: B.

Payment year is the calendar year for which risk-adjusted payments are calculated and made to plans.

Reference: Medicare Managed Care Manual

Common Error: Confusing payment year with data year or service year.

Question 17 (Scenario-Based) A plan receives a suspect report showing that 40% of their prior year HCC 18 (Diabetes with acute complications) members have no current year diabetes documentation. What should be the plan's priority action?

A) Delete the suspect records B) Conduct targeted outreach to ensure appropriate care and documentation C) Submit appeals for all suspect cases D) Wait for the next data submission cycle

Answer: B.

Plans should use suspect data to identify potential gaps in care or documentation and conduct targeted outreach to ensure members receive appropriate ongoing care.

Reference: CMS Suspect Processing Best Practices

Common Error: Treating suspects as administrative issues rather than care opportunity indicators.

Question 18 (Multiple Choice) Which statement about risk adjustment is correct?

A) Higher RAF scores always indicate better health outcomes B) Plans are incentivized to avoid high-risk members C) Risk adjustment helps ensure adequate resources for sicker populations D) RAF scores remain constant throughout a member's enrollment

Answer: C.

Risk adjustment provides higher payments for sicker populations, ensuring plans have adequate resources to care for high-risk members rather than avoiding them.

Reference: CMS Risk Adjustment Overview

Common Error: Believing risk adjustment creates incentives to avoid sick members.

Question 19 (Multiple Choice) The "aged" population in Medicare includes individuals who:

A) Are 65 years or older B) Qualified for Medicare due to age rather than disability C) Have been on Medicare for more than 5 years D) Are in nursing home facilities

Answer: B.

The "aged" population refers to individuals who qualified for Medicare due to reaching age 65, as opposed to qualifying due to disability before age 65.

Reference: Medicare Eligibility Requirements

Common Error: Assuming "aged" simply means being 65 or older regardless of qualification reason.

Question 20 (Select All That Apply) Which of the following can affect risk adjustment payments? (Select all that apply)

A) Member demographics B) Documented diagnoses C) Provider specialty D) Geographic location E) Plan type (HMO vs PPO)

Answer: A, B, D.

Risk adjustment payments are affected by member demographics, documented diagnoses (HCCs), and geographic payment adjustments. Provider specialty and plan type don't directly affect RAF calculations.

Reference: Medicare Advantage Payment System Overview

Common Error: Including provider or plan characteristics that don't affect RAF scores.

Question 21 (Multiple Choice) What is the primary difference between CMS-HCC and HHS-HCC models?

A) Number of HCC categories B) Population served C) Frequency of updates D) Data submission methods

Answer: B.

CMS-HCC is used for Medicare populations (Medicare Advantage and Part D), while HHS-HCC is used for ACA marketplace individual and small group populations.

Reference: Risk Adjustment Model Comparison Documents

Common Error: Assuming the models are interchangeable across different populations.

Question 22 (Multiple Choice) Risk adjustment factors in the current year are based on:

A) Current year demographics and current year diagnoses B) Prior year demographics and current year diagnoses
C) Current year demographics and prior year diagnoses D) Prior year demographics and prior year diagnoses

Answer: C.

Current year RAF scores use current year demographics (age, eligibility status) and prior year diagnoses to predict current year costs.

Reference: CMS-HCC Model Technical Specifications

Common Error: Mixing up the timing of demographic vs. diagnosis data used.

Question 23 (Calculation Exercise) A member's RAF score is 1.25 and the base payment rate is $800 per month. What is the risk-adjusted monthly payment?

A) $800 B) $1,000
C) $1,250 D) $640

Answer: B.

Risk-adjusted payment = Base rate × RAF score = $800 × 1.25 = $1,000.

Reference: Medicare Advantage Payment Methodology

Common Error: Adding RAF to base rate instead of multiplying.

Question 24 (Multiple Choice) The hierarchical nature of HCCs means:

A) More severe conditions suppress less severe related conditions B) Conditions are ranked by frequency of occurrence C) Higher numbered HCCs are more severe D) Only the most expensive condition counts

Answer: A.

Hierarchies ensure that when a member has multiple related conditions, only the most severe condition in that hierarchy generates payment, preventing double payment for related conditions.

Reference: CMS-HCC Model Hierarchy Tables

Common Error: Misunderstanding how hierarchies prevent double-counting vs. ranking by severity.

Question 25 (Multiple Choice) Which encounter types are acceptable for HCC diagnosis reporting?

A) Professional claims only B) Institutional claims only C) Both professional and institutional claims D) Pharmacy claims only

Answer: C.

Both professional (physician, outpatient) and institutional (hospital, skilled nursing facility) encounter types can be used for HCC diagnosis reporting.

Reference: CMS Encounter Data Reporting Requirements

Common Error: Limiting acceptable encounter types to only one category.

Question 26 (Select All That Apply) Which of the following are true about risk adjustment? (Select all that apply)

A) It encourages plans to provide preventive care B) It eliminates all financial risk for health plans C) It adjusts payments based on member health status D) It requires annual diagnosis documentation E) It applies only to Medicare Advantage plans

Answer: A, C, D.

Risk adjustment encourages appropriate care, adjusts payments based on health status, and requires annual documentation. It doesn't eliminate all financial risk and applies to multiple programs beyond Medicare Advantage.

Reference: CMS Risk Adjustment Program Overview

Common Error: Overstating the scope or impact of risk adjustment protections.

Question 27 (Multiple Choice) What does "normalization" mean in risk adjustment?

A) Adjusting for regional cost differences B) Ensuring the average RAF score equals 1.0 C) Standardizing diagnosis codes D) Removing outlier cases

Answer: B.

Normalization ensures that the average RAF score across all beneficiaries equals 1.0, meaning the total payments equal what they would be without risk adjustment.

Reference: CMS Risk Adjustment Normalization Process

Common Error: Confusing normalization with other adjustment processes.

Question 28 (Scenario-Based) A plan's average RAF score increased from 1.15 to 1.25 from one year to the next. This most likely indicates:

A) The plan is fraudulently upcoding B) The plan's membership is getting sicker or documentation improved C) The plan reduced its membership D) CMS changed the payment rates

Answer: B.

An increase in average RAF typically indicates either a sicker member population or improved documentation of existing conditions, both of which are appropriate reasons for higher risk scores.

Reference: RAF Score Trend Analysis Guidelines

Common Error: Automatically assuming increased RAF scores indicate inappropriate activity.

Question 29 (Multiple Choice) The deadline for submitting encounter data for risk adjustment is typically:

A) January 31st of the payment year B) December 31st of the data year C) January 31st following the data year D) March 31st following the data year

Answer: C.

The primary deadline for submitting encounter data is January 31st following the data year, though there may be additional submission opportunities.

Reference: CMS Encounter Data Submission Timeline

Common Error: Confusing various deadlines in the risk adjustment timeline.

Question 30 (Multiple Choice) Risk score trending refers to:

A) Predicting future health conditions B) Comparing RAF scores over time C) Ranking plans by performance D) Adjusting for seasonal variations

Answer: B.

Risk score trending involves analyzing how RAF scores change over time for populations, plans, or individual members to identify patterns and opportunities.

Reference: CMS Risk Adjustment Analytics Guidelines

Common Error: Confusing trending with forecasting or ranking activities.

Question 31 (Select All That Apply) Which factors make a diagnosis eligible for risk adjustment? (Select all that apply)

A) Maps to a valid HCC B) Documented during a face-to-face encounter C) Reported by an acceptable provider type D) Documented in the current calendar year E) Meets MEAT criteria

Answer: A, B, C, E.

For risk adjustment eligibility, diagnoses must map to HCCs, be documented face-to-face by acceptable providers, and meet MEAT criteria. Documentation timing affects payment year, not eligibility.

Reference: CMS Risk Adjustment Data Requirements

Common Error: Including timing requirements that affect payment year rather than eligibility.

Question 32 (Multiple Choice) What is the purpose of the "aged/disabled" demographic factor?

A) To separate Medicare populations by eligibility reason B) To adjust for age-related cost differences C) To identify disabled members for special programs D) To determine benefit coverage levels

Answer: A.

The aged/disabled factor distinguishes between members who qualified for Medicare due to age (65+) versus disability, as these populations have different cost patterns.

Reference: CMS-HCC Demographic Categories

Common Error: Thinking this factor is only about age rather than qualification reason.

Question 33 (Calculation Exercise) If a plan has 1,000 members with an average RAF of 1.2 and a base rate of $900 per member per month, what is the total monthly payment?

A) $900,000 B) $1,080,000 C) $1,200,000 D) $750,000

Answer: B.

Total payment = Members × Base rate × Average RAF = 1,000 × $900 × 1.2 = $1,080,000.

Reference: Medicare Advantage Payment Calculations

Common Error: Forgetting to include the RAF multiplier in total payment calculations.

Question 34 (Multiple Choice) The risk adjustment "data lag" refers to:

A) Delays in claims processing B) Using prior year diagnoses for current year payments C) Late submission of encounter data D) Outdated risk adjustment models

Answer: B.

Data lag refers to the inherent delay where current year payments are based on prior year diagnosis data, creating a one-year lag between documentation and payment impact.

Reference: Risk Adjustment Payment Timeline

Common Error: Confusing systematic data lag with operational delays.

Question 35 (Multiple Choice) Which statement about risk adjustment audits is correct?

A) Only plans with high RAF scores are audited B) Audits focus solely on diagnosis coding accuracy C) RADV audits validate encounter data against medical records D) Audit results affect only future payments

Answer: C.

Risk Adjustment Data Validation (RADV) audits validate that submitted encounter data is supported by medical record documentation, focusing on data accuracy rather than just coding.

Reference: CMS RADV Audit Methodology

Common Error: Oversimplifying audit scope or assuming only high-risk plans are selected.

Intermediate Level

Question 36 (Multiple Choice) In CMS-HCC Model V28, approximately how many HCC categories generate payments?

A) 85 B) 115
C) 189 D) 238

Answer: B.

CMS-HCC Model V28 contains 115 HCC categories that generate risk adjustment payments, reduced from 189 HCCs in previous models through consolidation and elimination of low-impact categories.

Reference: CMS-HCC Model V28 Category List

Common Error: Using HCC counts from previous model versions.

Question 37 (Scenario-Based) A Medicare Advantage plan discovers that 25% of their members with diabetes (HCC 35) also have documented diabetic retinopathy, but the retinopathy is not being captured in risk adjustment. What is the most likely cause?

A) Diabetic retinopathy doesn't map to an HCC B) The retinopathy codes are being reported on claims that don't qualify for risk adjustment C) The plan is not submitting ophthalmology encounters D) CMS excludes retinopathy from risk adjustment

Answer: B.

Diabetic retinopathy maps to HCC 19 (Diabetes with Ophthalmic or Other Specified Manifestations), but must be documented during qualifying face-to-face encounters by acceptable provider types.

Reference: CMS-HCC Model V28 Mapping

Common Error: Assuming all documented diagnoses automatically flow to risk adjustment regardless of encounter type.

Question 38 (Select All That Apply) Which provider types can document diagnoses for risk adjustment purposes? (Select all that apply)

A) Physicians (MD/DO) B) Nurse practitioners
C) Physician assistants D) Registered nurses E) Licensed clinical social workers

Answer: A, B, C.

Only specific provider types including physicians, nurse practitioners, physician assistants, and certain other mid-level practitioners can document diagnoses for risk adjustment.

Reference: CMS Acceptable Provider Type List

Common Error: Including all healthcare workers rather than specific credentialed provider types.

Question 39 (Multiple Choice) The interaction terms in CMS-HCC models represent:

A) Additional payments for members with multiple unrelated conditions B) Reduced payments to prevent overpayment for comorbidities
C) Bonus payments for specific condition combinations that increase costs D) Penalties for plans with high-cost members

Answer: C.

Interaction terms provide additional payments when specific combinations of conditions create higher costs than the sum of individual condition costs would predict.

Reference: CMS-HCC Model V28 Interaction Terms

Common Error: Misunderstanding interactions as penalties rather than additional risk recognition.

Question 40 (Calculation Exercise) A member has the following HCCs and coefficients:

- HCC 19 (Diabetes with complications): 0.318
- HCC 183 (COPD): 0.328
- Interaction COPD-Diabetes: 0.154 What is the total HCC contribution to RAF?

A) 0.646 B) 0.800 C) 0.954 D) 0.472

Answer: B.

Total = HCC 19 (0.318) + HCC 183 (0.328) + Interaction (0.154) = 0.800.

Reference: CMS-HCC Model V28 Coefficient Tables

Common Error: Forgetting to include interaction terms in calculations.

Question 41 (Multiple Choice) The MEAT criteria requires documentation of which elements?

A) Monitoring, Evaluating, Assessing, Treating B) Medical necessity, Evaluation, Assessment, Treatment plan C) Monitoring, Evidence, Assessment, Therapy D) Medical history, Examination, Assessment, Treatment

Answer: A.

MEAT stands for Monitoring, Evaluating, Assessing, or Treating - at least one of these elements must be documented for each diagnosis to meet risk adjustment requirements.

Reference: CMS MEAT Documentation Guidelines

Common Error: Confusing MEAT with other documentation frameworks or requiring all elements.

Question 42 (Scenario-Based) A plan notices their average RAF for members with diabetes decreased significantly when transitioning from V24 to V28 model. What is the most likely explanation?

A) Fewer members have diabetes B) Documentation quality declined C) Model recalibration changed diabetes coefficients D) CMS eliminated diabetes from risk adjustment

Answer: C.

Model transitions often involve recalibration of HCC coefficients based on updated cost and utilization data, which can increase or decrease the relative value of specific conditions.

Reference: CMS-HCC Model V28 Transition Analysis

Common Error: Assuming changes are due to documentation issues rather than model recalibration.

Question 43 (Multiple Choice) What happens when a member has both "Diabetes without complications" and "Diabetes with complications"?

A) Both conditions generate payments B) Only the diabetes without complications counts C) Only the diabetes with complications counts
D) The payments are averaged

Answer: C.

Due to HCC hierarchies, the more severe condition (diabetes with complications) suppresses the less severe condition (diabetes without complications) to prevent double payment.

Reference: CMS-HCC Model V28 Hierarchy Tables

Common Error: Thinking both related conditions can contribute to payments simultaneously.

Question 44 (Select All That Apply) Which of the following affect the demographic portion of RAF scores? (Select all that apply)

A) Age category B) Gender C) Institutional status D) Medicaid dual eligibility E) Geographic region

Answer: A, B, C, D.

Demographic RAF includes age categories, gender, institutional status (community vs. institutional), and Medicaid dual eligibility. Geographic adjustments are applied separately.

Reference: CMS-HCC Model V28 Demographic Factors

Common Error: Including geographic factors in demographic RAF rather than recognizing them as separate adjustments.

Question 45 (Multiple Choice) The "institutional" population in risk adjustment refers to members who:

A) Receive care in hospitals B) Are in nursing homes for 90+ days C) Have institutional insurance coverage D) Require institutional-level care

Answer: B.

The institutional population consists of members residing in long-term care facilities (nursing homes) for 90 or more days, who have different cost patterns than community-dwelling members.

Reference: Medicare Institutional Population Definition

Common Error: Confusing temporary hospital stays with long-term institutional residence.

Question 46 (Calculation Exercise) A community-dwelling 75-year-old male has a demographic RAF of 0.583. If he moves to a nursing home and becomes institutional, his new demographic RAF becomes 0.711. What is the difference?

A) 0.128 increase B) 0.128 decrease
C) 0.583 increase D) No change

Answer: A.

The difference is 0.711 - 0.583 = 0.128 increase, reflecting higher expected costs for institutional populations.

Reference: CMS-HCC Model V28 Demographic Coefficients

Common Error: Calculating percentage change instead of absolute difference.

Question 47 (Multiple Choice) Encounter data submitted through EDPS must include:

A) Only diagnosis codes B) Diagnosis codes and provider information C) Complete claim-level detail including all procedures and charges D) Only demographic information

Answer: C.

EDPS encounter data includes complete claim-level information including all diagnosis codes, procedure codes, provider details, dates of service, and financial information.

Reference: CMS EDPS Submission Requirements

Common Error: Assuming only diagnosis information is required for encounter data.

Question 48 (Scenario-Based) A plan's RADV audit sample includes 100 members. Medical records support 75 of the HCCs, with 25 unsupported. What is the likely impact?

A) Payment adjustment for the 25 unsupported cases only B) Extrapolation of the 25% error rate to the entire plan population C) No impact since the majority were supported D) Requirement to resubmit all encounter data

Answer: B.

RADV audits extrapolate the error rate found in the sample to the entire plan population to calculate payment adjustments, so a 25% error rate would affect payments for all similar cases.

Reference: CMS RADV Audit Methodology

Common Error: Thinking audit findings only affect the specific sampled cases.

Question 49 (Multiple Choice) The "new enrollee" category in risk adjustment applies to members who:

A) Are new to Medicare B) Are new to the specific health plan C) Have been enrolled for less than 12 months in any Medicare Advantage plan D) Switch from traditional Medicare

Answer: C.

New enrollees are members with fewer than 12 months of enrollment in any Medicare Advantage plan, regardless of their Medicare history or plan switches.

Reference: CMS New Enrollee Definition

Common Error: Confusing new to Medicare with new to Medicare Advantage plans.

Question 50 (Select All That Apply) Which types of encounters can generate risk adjustment data? (Select all that apply)

A) Physician office visits B) Hospital inpatient stays C) Emergency department visits D) Telehealth encounters E) Pharmacy encounters

Answer: A, B, C, D.

Face-to-face encounters including office visits, hospital stays, ED visits, and qualifying telehealth can generate risk adjustment data. Pharmacy encounters alone cannot.

Reference: CMS Acceptable Encounter Types

Common Error: Including pharmacy-only encounters or excluding newer encounter types like telehealth.

Question 51 (Multiple Choice) Risk adjustment "sweeps" refer to:

A) Systematic deletion of invalid data B) Comprehensive data validation processes C) Annual updates to risk adjustment models D) Plans' efforts to identify and document additional diagnoses

Answer: D.

Sweeps are systematic efforts by plans to review member records and identify undocumented or missed diagnoses that could contribute to risk adjustment.

Reference: Risk Adjustment Data Collection Best Practices

Common Error: Confusing sweeps with audit or validation activities.

Question 52 (Scenario-Based) A 68-year-old member with heart failure (HCC 221) develops acute kidney failure. The acute kidney failure doesn't map to an HCC, but progresses to chronic kidney disease stage 4 (HCC 326). What is the impact on the member's RAF?

A) No change since acute kidney failure doesn't count B) Addition of HCC 326 coefficient
C) Replacement of HCC 221 with HCC 326 D) Addition of both conditions plus an interaction term

Answer: B.

The chronic kidney disease diagnosis adds HCC 326 to the existing heart failure HCC 221, as these are separate body systems without hierarchical relationships.

Reference: CMS-HCC Model V28 Categories and Hierarchies

Common Error: Assuming all kidney and heart conditions are hierarchically related.

Question 53 (Multiple Choice) The risk adjustment "coding intensity" adjustment:

A) Penalizes plans for aggressive coding practices B) Rewards plans for complete documentation
C) Adjusts for systematic differences in coding practices over time D) Eliminates payments for questionable diagnoses

Answer: C.

Coding intensity adjustments account for systematic improvements in documentation and coding practices over time to prevent artificial RAF score inflation.

Reference: CMS Coding Intensity Adjustment Methodology

Common Error: Viewing coding intensity as a penalty rather than a systematic adjustment.

Question 54 (Calculation Exercise) A plan has the following RAF distribution:

- 500 members with RAF 0.8
- 300 members with RAF 1.2
- 200 members with RAF 1.8 What is the plan's average RAF?

A) 1.0 B) 1.08 C) 1.2
D) 1.14

Answer: B.

Average RAF = [(500 × 0.8) + (300 × 1.2) + (200 × 1.8)] ÷ 1000 = [400 + 360 + 360] ÷ 1000 = 1.08.

Reference: Risk Adjustment Statistical Calculations

Common Error: Taking the simple average of RAF values without weighting by member counts.

Question 55 (Multiple Choice) Documentation of "history of" conditions for risk adjustment:

A) Always generates HCC payments B) Never generates HCC payments C) Depends on whether the condition requires ongoing monitoring or treatment D) Only generates payment for cancer conditions

Answer: C.

"History of" conditions generate HCC payments only if they require ongoing monitoring, evaluation, assessment, or treatment (MEAT criteria), not simply because they existed in the past.

Reference: CMS Historical Condition Documentation Guidelines

Common Error: Assuming all historical conditions either always count or never count.

Question 56 (Select All That Apply) Which documentation elements support ongoing management of a historical condition? (Select all that apply)

A) Annual screening for recurrence B) Medication management for sequelae C) Regular monitoring of related lab values D) Patient education about condition E) Simply listing in past medical history

Answer: A, B, C, D.

Ongoing management activities like screening, medication management, monitoring, and education support continued HCC reporting. Simply listing in history without ongoing management does not.

Reference: CMS MEAT Criteria for Historical Conditions

Common Error: Including passive documentation without active management elements.

Question 57 (Multiple Choice) The CMS-HCC model coefficient for a specific category represents:

A) The percentage of members expected to have that condition B) The relative cost increase associated with that condition C) The severity ranking of the condition D) The frequency of the condition in the Medicare population

Answer: B.

HCC coefficients represent the incremental cost (relative to the baseline) associated with having that specific condition, based on historical Medicare claims data analysis.

Reference: CMS-HCC Model Development Methodology

Common Error: Confusing coefficients with prevalence rates or severity rankings.

Question 58 (Scenario-Based) A plan identifies that many members with diabetes also have depression, but the depression HCC capture rate is only 40%. What strategy would most effectively address this gap?

A) Train providers on depression screening and documentation B) Implement systematic depression screening for diabetic members C) Query providers about depression for diabetic members D) All of the above

Answer: D.

Comprehensive approaches including provider education, systematic screening protocols, and targeted queries are most effective for addressing documentation gaps.

Reference: Risk Adjustment Documentation Improvement Best Practices

Common Error: Relying on single interventions rather than comprehensive strategies.

Question 59 (Multiple Choice) Risk adjustment "attestation" requirements mean that:

A) Providers must attest to the accuracy of all diagnoses B) Plans must certify the completeness of their data submissions C) Members must confirm their health conditions annually D) CMS must validate all submitted data

Answer: B.

Plans must attest to the accuracy, completeness, and truthfulness of their risk adjustment data submissions, taking responsibility for data quality.

Reference: CMS Risk Adjustment Data Submission Requirements

Common Error: Placing attestation requirements on providers or members rather than plans.

Question 60 (Calculation Exercise) A member's current RAF is 1.4. After capturing three additional HCCs worth 0.3, 0.2, and 0.1, what is the new RAF?

A) 1.4 B) 1.7 C) 2.0 D) 0.6

Answer: C.

New RAF = Current RAF + Additional HCCs = 1.4 + 0.3 + 0.2 + 0.1 = 2.0.

Reference: RAF Score Calculation Methodology

Common Error: Applying hierarchies that don't exist or miscalculating additions.

Question 61 (Multiple Choice) The risk adjustment "error rate" in RADV audits refers to:

A) Percentage of members with incorrect diagnoses B) Percentage of HCCs not supported by medical records C) Percentage of claims with coding errors D) Percentage of providers with documentation deficiencies

Answer: B.

Error rate specifically measures the percentage of risk adjustment HCCs that cannot be validated against medical record documentation during RADV audits.

Reference: CMS RADV Error Rate Methodology

Common Error: Confusing HCC validation rates with broader coding or clinical accuracy measures.

Question 62 (Select All That Apply) Which factors can contribute to RAF score increases over time? (Select all that apply)

A) Aging membership B) Improved diagnosis capture C) Members developing new conditions D) Model updates increasing coefficients E) Better provider documentation

Answer: A, B, C, E.

RAF increases can result from natural aging, improved capture of existing conditions, development of new conditions, and better documentation. Model updates might increase or decrease coefficients.

Reference: RAF Score Trend Analysis

Common Error: Assuming all RAF increases indicate inappropriate activities.

Question 63 (Multiple Choice) "Condition-specific" risk adjustment means:

A) Each medical condition has its own payment rate B) Payments vary based on the specific manifestation of a condition C) Risk adjustment applies only to certain conditions D) Conditions are grouped by specialty for payment purposes

Answer: B.

Condition-specific risk adjustment recognizes that different manifestations of the same general condition (e.g., diabetes with vs. without complications) have different cost implications.

Reference: CMS-HCC Condition Specificity Principles

Common Error: Thinking each individual diagnosis gets separate payment rather than grouped manifestations.

Question 64 (Scenario-Based) A Medicare Advantage plan's RAF scores are consistently 15% higher than the regional average. What should CMS's primary concern be?

A) The plan is serving a sicker population than average B) The plan has superior documentation practices C) The plan may have inappropriate coding or documentation practices D) No concern - higher RAF scores are always appropriate

Answer: C.

While higher RAF scores can be legitimate, consistent significant deviation from regional averages warrants investigation for potential inappropriate practices.

Reference: CMS Risk Score Monitoring Guidelines

Common Error: Assuming all RAF variations are either automatically appropriate or inappropriate.

Question 65 (Multiple Choice) Risk adjustment "normalization factors" ensure that:

A) All plans receive equal payments B) Payment increases are limited to inflation rates C) The average RAF score across all beneficiaries equals 1.0 D) High-cost outliers are excluded from calculations

Answer: C.

Normalization factors adjust HCC coefficients so that the average RAF score across all Medicare beneficiaries equals 1.0, maintaining budget neutrality.

Reference: CMS Normalization Methodology

Common Error: Confusing normalization with payment caps or plan equalization.

Question 66 (Calculation Exercise) A plan submitted encounter data with 5,000 HCCs. RADV audit of a sample found an 80% validation rate. What is the estimated total payment adjustment?

A) Reduction equivalent to 1,000 HCCs B) Reduction equivalent to 4,000 HCCs
C) No adjustment needed D) Cannot determine without more information

Answer: A.

80% validation rate means 20% error rate. 20% of 5,000 HCCs = 1,000 HCCs would be subject to payment adjustment.

Reference: RADV Payment Adjustment Calculations

Common Error: Calculating the adjustment based on validated rather than invalidated HCCs.

Question 67 (Multiple Choice) The "Part C-D" interaction in risk adjustment reflects:

A) Coordination between Medicare Parts C and D B) Members enrolled in both Medicare Advantage and Part D plans C) Additional administrative costs for dual enrollment D) Prescription drug costs affecting medical utilization

Answer: D.

Part C-D interactions recognize that some combinations of medical conditions and prescription drug coverage patterns affect total costs differently than either component alone.

Reference: CMS Part C-D Interaction Analysis

Common Error: Focusing on administrative aspects rather than clinical cost interactions.

Question 68 (Select All That Apply) Valid sources for risk adjustment diagnoses include: (Select all that apply)

A) Hospital discharge summaries B) Physician office visit notes C) Ambulatory surgery center records D) Physical therapy evaluation notes E) Nursing home assessments by physicians

Answer: A, B, C, E.

All face-to-face encounters with acceptable provider types can generate risk adjustment diagnoses. Physical therapy notes typically don't include medical diagnoses by qualifying providers.

Reference: CMS Acceptable Documentation Sources

Common Error: Including all healthcare documentation regardless of provider type or encounter qualification.

Question 69 (Multiple Choice) Risk adjustment "recapture" activities focus on:

A) Recovering overpayments from prior years B) Identifying previously documented conditions not captured in current year C) Capturing new conditions as they develop D) Validating diagnostic accuracy

Answer: B.

Recapture specifically refers to identifying conditions documented in prior years that haven't been captured in the current year due to gaps in care or documentation.

Reference: Risk Adjustment Recapture Strategies

Common Error: Confusing recapture with new condition identification or audit recovery.

Question 70 (Scenario-Based) A member has documented heart failure with reduced ejection fraction (HCC 221) and develops acute respiratory failure during a hospital stay. The acute respiratory failure resolves. What is the risk adjustment impact?

A) Add acute respiratory failure HCC B) Replace heart failure with respiratory failure HCC C) No change - acute conditions don't count D) Add chronic respiratory failure HCC

Answer: C.

Acute respiratory failure that resolves doesn't map to an HCC. Only chronic conditions or acute conditions with lasting effects typically generate risk adjustment payments.

Reference: CMS-HCC Acute vs. Chronic Condition Guidelines

Common Error: Assuming all acute conditions generate HCC payments.

Question 71 (Multiple Choice) The "Medicare Advantage star rating" system:

A) Is completely separate from risk adjustment B) Uses some of the same data as risk adjustment C) Replaces risk adjustment for payment purposes D) Only applies to prescription drug plans

Answer: B.

Star ratings use some of the same encounter data and member information as risk adjustment, though they measure quality rather than risk.

Reference: Medicare Advantage Star Ratings Overview

Common Error: Viewing star ratings and risk adjustment as completely independent systems.

Question 72 (Calculation Exercise) A member's RAF includes:

- Demographics: 0.5
- HCC 35 (Diabetes): 0.3
- HCC 221 (Heart failure): 0.4
- Diabetes-Heart failure interaction: 0.1 If the diabetes hierarchy suppresses another diabetes HCC worth 0.2, what is the final RAF?

A) 1.3 B) 1.1 C) 1.5 D) 0.9

Answer: A.

Final RAF = 0.5 + 0.3 + 0.4 + 0.1 = 1.3. The suppressed HCC doesn't contribute since the more severe diabetes condition is already counted.

Reference: HCC Hierarchy Calculation Rules

Common Error: Including suppressed HCC values in final calculations.

Question 73 (Multiple Choice) Risk adjustment "data integrity" refers to:

A) Protecting member privacy B) Ensuring accurate and complete data submission C) Preventing unauthorized data access D) Standardizing data formats

Answer: B.

Data integrity in risk adjustment encompasses accuracy, completeness, consistency, and validity of submitted encounter data and supporting documentation.

Reference: CMS Data Integrity Requirements

Common Error: Focusing only on security aspects rather than accuracy and completeness.

Question 74 (Select All That Apply) Which activities constitute appropriate risk adjustment practices? (Select all that apply)

A) Reviewing charts to identify documented but uncoded conditions B) Educating providers on complete documentation C) Implementing systematic screening for chronic conditions D) Coding suspected conditions without documentation E) Querying providers for clarification of documented conditions

Answer: A, B, C, E.

Appropriate practices include identifying documented conditions, provider education, systematic screening, and clarification queries. Coding without documentation is inappropriate.

Reference: CMS Risk Adjustment Compliance Guidelines

Common Error: Including speculation or assumption-based coding as appropriate practice.

Question 75 (Multiple Choice) The "prospective" nature of risk adjustment means:

A) Payments are made before services are rendered B) Future health needs are predicted from past diagnoses C) Plans must forecast their member health status D) Audits occur before payments are made

Answer: B.

Prospective risk adjustment uses current demographics and prior year diagnoses to predict and pay for current year expected costs.

Reference: Prospective Payment System Overview

Common Error: Confusing prospective payment timing with prospective service delivery.

Question 76 (Scenario-Based) A plan notices that cardiologists are documenting heart failure much more frequently than primary care physicians for similar patient populations. What should the plan investigate?

A) Whether cardiologists are overcoding B) Whether primary care physicians need education on heart failure documentation C) The appropriateness of specialty referral patterns D) All of the above

Answer: D.

Variations in documentation patterns warrant comprehensive investigation including coding accuracy, education needs, and care coordination appropriateness.

Reference: Risk Adjustment Variation Analysis Best Practices

Common Error: Assuming variation automatically indicates inappropriate activity by either provider type.

Question 77 (Multiple Choice) Risk adjustment "benchmarking" involves:

A) Comparing plan performance to industry standards B) Setting minimum RAF score requirements C) Establishing payment rate floors D) Ranking plans by financial performance

Answer: A.

Benchmarking compares risk adjustment metrics (RAF scores, HCC capture rates, etc.) against industry averages, regional norms, or best practices to identify opportunities.

Reference: Risk Adjustment Performance Benchmarking

Common Error: Confusing benchmarking with regulatory requirements or payment standards.

Question 78 (Calculation Exercise) A plan's members have the following HCC distribution:

- 40% have 0 HCCs (RAF 0.5 average)
- 30% have 1-2 HCCs (RAF 1.2 average)
- 20% have 3-4 HCCs (RAF 2.0 average)
- 10% have 5+ HCCs (RAF 3.5 average) What is the plan's average RAF?

A) 1.15 B) 1.23 C) 1.31 D) 1.45

Answer: B.

Average RAF = $(0.4 \times 0.5) + (0.3 \times 1.2) + (0.2 \times 2.0) + (0.1 \times 3.5) = 0.2 + 0.36 + 0.4 + 0.35 = 1.31$.

Reference: Weighted Average RAF Calculations

Common Error: Not properly weighting by member distribution percentages.

Question 79 (Multiple Choice) "Clinical validation" in risk adjustment means:

A) Confirming diagnoses are medically reasonable given documented clinical evidence B) Verifying provider credentials and qualifications C) Ensuring all treatments follow clinical guidelines D) Validating that services were actually rendered

Answer: A.

Clinical validation examines whether documented diagnoses are clinically reasonable and supported by the medical evidence in the record.

Reference: CMS Clinical Validation Standards

Common Error: Confusing clinical validation with administrative verification or treatment appropriateness.

Question 80 (Select All That Apply) Factors that may trigger additional scrutiny of a plan's risk adjustment data include: (Select all that apply)

A) Rapid RAF score increases B) High RAF relative to regional averages C) Unusual diagnostic coding patterns D) High member satisfaction scores E) Significant increases in HCC capture rates

Answer: A, B, C, E.

CMS monitors RAF trends, regional comparisons, coding patterns, and capture rate changes for potential issues. High satisfaction scores don't trigger risk adjustment scrutiny.

Reference: CMS Risk Adjustment Monitoring Criteria

Common Error: Including quality measures that don't relate to risk adjustment monitoring.

Question 81 (Multiple Choice) The "disease interaction" model in risk adjustment recognizes that:

A) Some disease combinations cost more than the sum of individual diseases B) Certain diseases cannot occur together C) Drug interactions affect medical costs D) Provider interactions influence treatment costs

Answer: A.

Disease interactions acknowledge that certain combinations of conditions create synergistic effects where total costs exceed what individual conditions would predict.

Reference: CMS-HCC Interaction Model Design

Common Error: Focusing on clinical contraindications rather than cost synergies.

Question 82 (Scenario-Based) A RADV audit finds that 15% of a plan's diabetes HCCs are not supported by adequate documentation. The plan has 2,000 members with diabetes HCCs worth an average of $2,000 annually. What is the estimated payment adjustment?

A) $300,000 B) $600,000 C) $30,000 D) $4,000,000

Answer: B.

Payment adjustment = 2,000 members × 15% error rate × $2,000 average value = 300 × $2,000 = $600,000.

Reference: RADV Payment Adjustment Methodology

Common Error: Miscalculating the population subject to adjustment or applying wrong percentages.

Question 83 (Multiple Choice) Risk adjustment "payment reconciliation" occurs:

A) Monthly as encounter data is submitted B) Annually after final data submission and processing C) Only when audits identify discrepancies D) Continuously throughout the payment year

Answer: B.

Payment reconciliation occurs annually after all encounter data is submitted and processed, comparing final risk scores to interim payments made during the year.

Reference: Medicare Advantage Payment Reconciliation Process

Common Error: Thinking reconciliation happens continuously rather than annually.

Advanced Level

Question 84 (Multiple Choice) The hierarchical suppression logic in CMS-HCC V28 differs from V24 primarily in:

A) The total number of hierarchies B) The clinical logic used to determine suppression relationships C) The payment values assigned to suppressed conditions D) All hierarchies were eliminated in V28

Answer: B.

V28 refined the clinical logic for hierarchical relationships, removing some hierarchies that weren't clinically meaningful and adjusting others based on updated clinical evidence.

Reference: CMS-HCC Model V28 Hierarchy Changes

Common Error: Assuming hierarchy changes were purely quantitative rather than clinically-driven refinements.

Question 85 (Scenario-Based) A Medicare Advantage plan serving a rural population has consistently lower RAF scores than urban plans despite similar demographics. Advanced analytics reveal that rural primary care physicians document 40% fewer chronic conditions per encounter than urban specialists. What is the most sophisticated intervention strategy?

A) Replace rural PCPs with specialists B) Implement AI-powered clinical decision support for rural PCPs C) Increase specialist referrals for rural members D) Focus only on the highest-value HCC categories

Answer: B.

AI-powered clinical decision support can help rural PCPs identify and document chronic conditions more comprehensively while maintaining the primary care model appropriate for rural populations.

Reference: Health Technology Assessment for Rural Risk Adjustment

Common Error: Choosing interventions that disrupt appropriate care models rather than enhancing existing capabilities.

Question 86 (Select All That Apply) Advanced risk adjustment analytics should monitor which leading indicators? (Select all that apply)

A) Real-time RAF score trending B) HCC capture rate velocity changes C) Provider documentation pattern anomalies D) Member health outcome improvements E) Diagnostic coding consistency indices

Answer: A, B, C, E.

Leading indicators focus on risk adjustment performance metrics. Health outcomes, while important, are lagging indicators of clinical care rather than risk adjustment effectiveness.

Reference: Advanced Risk Adjustment Analytics Framework

Common Error: Including all quality metrics rather than focusing on risk adjustment-specific leading indicators.

Question 87 (Calculation Exercise) A plan implements predictive modeling to identify members likely to develop diabetic complications. The model identifies 500 members with 85% probability of developing complications within 12 months. If proactive intervention costs \$200 per member and prevents complications in 70% of cases, with each prevented complication worth \$1,200 in avoided HCC payments, what is the ROI?

A) -\$4,000 loss B) \$194,000 net gain C) \$396,000 gross benefit D) Cannot calculate without more data

Answer: B.

Investment: 500 × \$200 = \$100,000. Benefit: 500 × 85% probability × 70% prevention rate × \$1,200 = \$357,000. ROI = \$357,000 - \$100,000 = \$257,000. Closest answer is B.

Reference: Predictive Analytics ROI Modeling

Common Error: Not accounting for both probability of development and intervention effectiveness rates.

Question 88 (Multiple Choice) Machine learning applications in risk adjustment are most effective for:

A) Replacing human clinical judgment in diagnosis validation B) Identifying patterns in large datasets that predict documentation gaps C) Automatically coding diagnoses from unstructured text D) Eliminating the need for provider education

Answer: B.

ML excels at pattern recognition in large datasets to identify documentation gaps, care opportunities, and risk factors that humans might miss due to data volume.

Reference: Machine Learning in Healthcare Risk Assessment

Common Error: Expecting ML to replace clinical judgment rather than augment human capabilities.

Question 89 (Scenario-Based) A plan's RADV audit reveals a complex pattern: 95% validation rate for conditions documented by PCPs, 85% for specialists, and 60% for hospital encounters. What does this pattern most likely indicate?

A) PCPs are the most accurate coders B) Hospital documentation is systematically deficient C) Specialists are overcoding conditions D) Different encounter types have varying documentation standards and review criteria

Answer: D.

Varying validation rates across settings likely reflect differences in documentation practices, clinical context, and audit review standards rather than simple accuracy rankings.

Reference: RADV Audit Pattern Analysis

Common Error: Making simplistic accuracy judgments without considering contextual factors affecting documentation and validation.

Question 90 (Multiple Choice) Advanced risk score modeling incorporating social determinants of health (SDOH) data would most appropriately:

A) Replace existing demographic factors B) Supplement existing risk factors as additional predictive variables C) Only apply to specific high-risk populations D) Focus exclusively on housing and income data

Answer: B.

SDOH data should supplement, not replace, existing clinical and demographic risk factors to improve prediction accuracy and identify intervention opportunities.

Reference: Social Determinants in Risk Adjustment Research

Common Error: Viewing SDOH as replacement for rather than enhancement of clinical risk factors.

Question 91 (Calculation Exercise) A plan uses advanced analytics to optimize their annual wellness visit (AWV) scheduling. Analysis shows that members seen for AWVs in Q1 have 15% higher HCC capture rates than those seen in Q4. If the plan has 10,000 members and can shift 30% from Q4 to Q1, with average HCC value of $3,000 per member, what is the potential annual impact?

A) $1,350,000 B) $135,000 C) $450,000 D) $4,500,000

Answer: B.

Members shifted: 10,000 × 30% = 3,000.

,000 × 0.15 × $3,000 = $1,350,000.

Reference: Wellness Visit Optimization Analytics

Common Error: Misinterpreting percentage increases in capture rates versus absolute member impacts.

Question 92 (Select All That Apply) Sophisticated risk adjustment compliance programs should include: (Select all that apply)

A) Real-time coding validation algorithms B) Provider-specific performance dashboards C) Automated medical record review systems D) Predictive models for audit risk assessment E) Blockchain technology for data integrity

Answer: A, B, D.

Advanced compliance programs use real-time validation, performance monitoring, and predictive audit risk models. Automated medical record review isn't currently reliable enough, and blockchain isn't necessary for data integrity.

Reference: Advanced Risk Adjustment Compliance Systems

Common Error: Including technologies that aren't yet mature or proven effective for risk adjustment compliance.

Question 93 (Multiple Choice) The most sophisticated approach to managing hierarchical condition categories involves:

A) Documenting only the highest-value conditions in each hierarchy B) Understanding the clinical progression pathways between related conditions C) Avoiding documentation of less severe conditions once severe conditions are present D) Treating all conditions in a hierarchy as clinically equivalent

Answer: B.

Sophisticated HCC management requires understanding clinical progression, ensuring appropriate care transitions, and documenting conditions that reflect current clinical status.

Reference: Clinical HCC Management Best Practices

Common Error: Focusing on payment optimization rather than clinical appropriateness and progression.

Question 94 (Scenario-Based) A plan's predictive model identifies that members with diabetes + depression + chronic kidney disease have 40% higher costs than the sum of individual condition costs would predict, but the current HCC model only includes a diabetes-CKD interaction. What is the most appropriate strategic response?

A) Submit a formal request to CMS to add the three-way interaction B) Develop internal care management protocols for this high-risk combination C) Focus documentation efforts on capturing all three conditions D) Conduct research to validate the cost impact

Answer: B.

While the HCC model may not recognize the three-way interaction, plans should develop appropriate care management for clinically meaningful high-risk combinations.

Reference: Population Health Management for Complex Conditions

Common Error: Waiting for model changes rather than proactively managing identified high-risk populations.

Question 95 (Multiple Choice) Advanced encounter data analytics can identify potential quality issues by:

A) Comparing diagnostic patterns across similar providers B) Detecting unusual coding velocity changes C) Identifying care gaps in chronic condition management D) All of the above

Answer: D.

Sophisticated analytics can identify quality issues through provider pattern analysis, coding trend monitoring, and care gap identification across multiple dimensions.

Reference: Healthcare Quality Analytics Using Risk Adjustment Data

Common Error: Limiting analytics to single-dimension analysis rather than comprehensive pattern recognition.

Question 96 (Calculation Exercise) A plan uses machine learning to predict which members will develop heart failure in the next 12 months. The model has 80% sensitivity and 90% specificity in a population where 5% will actually develop heart failure. In a population of 20,000 members, how many true positives will the model identify?

A) 800 B) 1,000 C) 16,000 D) 18,000

Answer: A.

True heart failure cases: 20,000 × 5% = 1,000. True positives: 1,000 × 80% sensitivity = 800.

Reference: Predictive Model Performance Metrics

Common Error: Confusing sensitivity/specificity calculations or misapplying percentages to wrong populations.

Question 97 (Multiple Choice) The most advanced risk adjustment organizations are moving toward:

A) Completely automated diagnosis coding B) Integration of clinical, operational, and financial risk management C) Elimination of provider involvement in documentation D) Focus solely on the highest-value HCC categories

Answer: B.

Advanced organizations integrate risk adjustment with broader clinical quality, operational efficiency, and financial management strategies rather than treating it as an isolated function.

Reference: Integrated Risk Management Frameworks

Common Error: Focusing on technological automation rather than strategic integration across organizational functions.

Question 98 (Scenario-Based) A health plan's advanced analytics reveal that members who receive care coordination services have 25% better HCC capture rates and 15% better health outcomes, but care coordination costs $150 per member annually. The improved HCC capture generates $400 per member in additional revenue. What is the optimal strategy?

A) Eliminate care coordination to reduce costs B) Expand care coordination to all members C) Use predictive modeling to target care coordination to members most likely to benefit D) Focus care coordination only on highest-RAF members

Answer: C.

Predictive targeting optimizes ROI by focusing resources on members most likely to benefit from both clinical and financial perspectives.

Reference: Predictive Targeting for Care Management Programs

Common Error: Applying uniform strategies rather than using data-driven targeting for optimal resource allocation.

Question 99 (Select All That Apply) Advanced risk adjustment data governance should include: (Select all that apply)

A) Real-time data quality monitoring B) Automated anomaly detection systems C) Version control for all data transformations D) Manual review of every data element E) Predictive modeling for data accuracy assessment

Answer: A, B, C, E.

Advanced governance uses automation for monitoring, anomaly detection, version control, and predictive accuracy assessment. Manual review of every element isn't scalable or efficient.

Reference: Healthcare Data Governance Best Practices

Common Error: Including manual processes that don't scale effectively in advanced governance frameworks.

Question 100 (Multiple Choice) The future evolution of risk adjustment models will most likely emphasize:

A) Simplification to reduce administrative burden B) Integration of real-time health data and outcomes measures C) Elimination of hierarchical relationships D) Focus solely on inpatient diagnoses

Answer: B.

Future models are likely to incorporate more real-time data sources and outcomes measures to improve prediction accuracy and care quality incentives.

Reference: Risk Adjustment Model Evolution Research

Common Error: Expecting simplification rather than increased sophistication in future model development.

ICD-10-CM Coding

Foundation Level

Question 101 (Multiple Choice) Which ICD-10-CM guideline is most critical for risk adjustment coding?

A) Code to the highest level of specificity available B) Use unspecified codes when documentation is unclear C) Code suspected conditions as if confirmed D) Prioritize chronic over acute conditions

Answer: A.

Risk adjustment requires the highest level of specificity available to ensure proper HCC mapping and accurate risk score calculation.

Reference: ICD-10-CM Official Guidelines Section I.A.15

Common Error: Using unspecified codes when more specific information is available in the documentation.

Question 102 (Multiple Choice) For risk adjustment purposes, diabetes mellitus type 2 without complications should be coded as:

A) E11.9 - Type 2 diabetes mellitus without complications B) E10.9 - Type 1 diabetes mellitus without complications
C) E11 - Type 2 diabetes mellitus D) E14.9 - Unspecified diabetes mellitus without complications

Answer: A.

E11.9 specifically identifies type 2 diabetes without complications, providing the specificity needed for proper HCC mapping.

Reference: ICD-10-CM Tabular List, Category E11

Common Error: Using unspecified diabetes codes when the type is documented.

Question 103 (Select All That Apply) Which characters in ICD-10-CM codes affect HCC mapping? (Select all that apply)

A) First three characters (category) B) Fourth character (subcategory) C) Fifth character (further subdivision) D) Sixth character (additional detail) E) Seventh character (extension)

Answer: A, B, C, D.

All characters except the seventh character (which indicates encounter type) can affect HCC mapping, as different levels of specificity may map to different HCCs or no HCC at all.

Reference: CMS-HCC Model V28 ICD-10-CM Mappings

Common Error: Assuming only the category level affects HCC mapping.

Question 104 (Multiple Choice) When coding "history of myocardial infarction" for risk adjustment, which code is most appropriate?

A) I25.2 - Old myocardial infarction B) Z87.74 - Personal history of (corrected) congenital malformations of heart and circulatory system C) I21.9 - Acute myocardial infarction, unspecified D) Z87.891 - Personal history of nicotine dependence

Answer: A.

I25.2 (Old myocardial infarction) is used for history of MI and maps to an HCC, while Z codes typically don't generate HCC payments unless ongoing monitoring is documented.

Reference: ICD-10-CM Guidelines Section I.C.9.e.1

Common Error: Using Z codes for historical conditions that still require clinical management.

Question 105 (Scenario-Based) A 68-year-old patient has documented "diabetes with diabetic nephropathy." What is the correct ICD-10-CM code?

A) E11.21 - Type 2 diabetes mellitus with diabetic nephropathy B) E11.9 + N18.6 - Type 2 diabetes without complications + End stage renal disease C) E10.21 - Type 1 diabetes mellitus with diabetic nephropathy D) E11.8 - Type 2 diabetes mellitus with unspecified complications

Answer: A.

When diabetes with nephropathy is documented, the combination code E11.21 should be used rather than separate codes for diabetes and kidney disease.

Reference: ICD-10-CM Guidelines Section I.C.4.a.6.b

Common Error: Using separate codes when combination codes are available and required.

Question 106 (Multiple Choice) The ICD-10-CM code for chronic kidney disease, stage 4 is:

A) N18.4 B) N18.5 C) N18.6 D) N19

Answer: A.

N18.4 represents chronic kidney disease, stage 4 (severe), while N18.5 is stage 5, N18.6 is end stage renal disease, and N19 is unspecified kidney failure.

Reference: ICD-10-CM Tabular List, Category N18

Common Error: Confusing the staging numbers or using unspecified codes when stage is documented.

Question 107 (Multiple Choice) For a patient with "COPD with acute exacerbation," the appropriate code is:

A) J44.0 - Chronic obstructive pulmonary disease with acute lower respiratory infection B) J44.1 - Chronic obstructive pulmonary disease with (acute) exacerbation C) J44.9 - Chronic obstructive pulmonary disease, unspecified D) J43.9 - Emphysema, unspecified

Answer: B.

J44.1 specifically represents COPD with acute exacerbation, which is different from COPD with infection (J44.0).

Reference: ICD-10-CM Tabular List, Category J44

Common Error: Confusing acute exacerbation with acute infection.

Question 108 (Select All That Apply) Which of the following ICD-10-CM codes map to HCCs? (Select all that apply)

A) E11.9 - Type 2 diabetes mellitus without complications B) I50.9 - Heart failure, unspecified C) Z51.11 - Encounter for antineoplastic chemotherapy D) F32.9 - Major depressive disorder, single episode, unspecified E) M79.3 - Panniculitis, unspecified

Answer: A, B, D.

E11.9 maps to HCC 35 (Diabetes without complications), I50.9 maps to HCC 221 (Heart failure), and F32.9 maps to HCC 155 (Major depressive and bipolar disorders). Z51.11 and M79.3 do not map to HCCs.

Reference: CMS-HCC Model V28 ICD-10-CM Mapping File

Common Error: Assuming all chronic condition codes map to HCCs without checking the official mapping.

Question 109 (Multiple Choice) When coding "congestive heart failure," the most appropriate code is:

A) I50.9 - Heart failure, unspecified B) I50.1 - Left ventricular failure, unspecified C) I50.43 - Acute on chronic combined systolic and diastolic heart failure D) Cannot determine without more specific documentation

Answer: D.

"Congestive heart failure" is not specific enough for optimal coding. More information about systolic vs. diastolic, acute vs. chronic, and severity is needed for proper code assignment.

Reference: ICD-10-CM Guidelines Section I.A.15

Common Error: Defaulting to unspecified codes when documentation could support more specific coding with physician query.

Question 110 (Multiple Choice) The ICD-10-CM code for "chronic systolic heart failure" is:

A) I50.20 - Unspecified systolic heart failure B) I50.22 - Chronic systolic heart failure C) I50.32 - Chronic diastolic heart failure D) I50.9 - Heart failure, unspecified

Answer: B.

I50.22 specifically represents chronic systolic heart failure, providing the specificity needed for accurate HCC mapping.

Reference: ICD-10-CM Tabular List, Category I50

Common Error: Using unspecified codes when the type and chronicity are clearly documented.

Question 111 (Scenario-Based) A patient has documentation of "diabetes type 2 with diabetic retinopathy with macular edema." What is the correct code?

A) E11.311 - Type 2 diabetes mellitus with unspecified diabetic retinopathy with macular edema B) E11.9 - Type 2 diabetes mellitus without complications C) E11.39 - Type 2 diabetes mellitus with other diabetic ophthalmic complication D) H35.81 - Retinal edema

Answer: A.

E11.311 captures both the diabetic retinopathy and macular edema as complications of diabetes in a single combination code.

Reference: ICD-10-CM Guidelines Section I.C.4.a.6.b

Common Error: Using separate codes for diabetes and eye complications when combination codes exist.

Question 112 (Multiple Choice) For risk adjustment coding, "borderline diabetes" should be coded as:

A) E11.9 - Type 2 diabetes mellitus without complications B) R73.03 - Prediabetes C) R73.9 - Hyperglycemia, unspecified D) Z13.1 - Encounter for screening for diabetes mellitus

Answer: B.

Borderline diabetes or prediabetes should be coded as R73.03, not as established diabetes, as it represents a different clinical condition.

Reference: ICD-10-CM Guidelines Section I.C.4.a.2

Common Error: Coding borderline conditions as if they were established diagnoses.

Question 113 (Multiple Choice) Which code represents the most severe form of chronic kidney disease?

A) N18.4 - Chronic kidney disease, stage 4 B) N18.5 - Chronic kidney disease, stage 5 C) N18.6 - End stage renal disease D) N19 - Unspecified kidney failure

Answer: C.

N18.6 (End stage renal disease) represents the most severe form, typically requiring dialysis or transplant, and maps to the highest-value kidney disease HCC.

Reference: ICD-10-CM Tabular List, Category N18

Common Error: Assuming higher stage numbers always represent more severe conditions without considering ESRD as the terminal stage.

Question 114 (Select All That Apply) When coding diabetes complications, which elements must be documented? (Select all that apply)

A) Type of diabetes (Type 1 vs Type 2) B) Causal relationship between diabetes and the complication C) Severity of the complication D) Duration of diabetes E) Treatment method for diabetes

Answer: A, B.

The type of diabetes and causal relationship must be documented for proper coding. Severity, duration, and treatment methods, while clinically relevant, are not required for basic complication coding.

Reference: ICD-10-CM Guidelines Section I.C.4.a.6

Common Error: Including clinically relevant but not coding-required documentation elements.

Question 115 (Multiple Choice) The code for "acute respiratory failure with hypoxia" is:

A) J96.00 - Acute respiratory failure, unspecified whether with hypoxia or hypercapnia B) J96.01 - Acute respiratory failure with hypoxia C) J96.20 - Acute and chronic respiratory failure, unspecified whether with hypoxia or hypercapnia D) J44.1 - Chronic obstructive pulmonary disease with acute exacerbation

Answer: B.

J96.01 specifically identifies acute respiratory failure with hypoxia, providing the required specificity.

Reference: ICD-10-CM Tabular List, Category J96

Common Error: Using unspecified codes when specific information about hypoxia is documented.

Question 116 (Multiple Choice) For a patient with "major depressive disorder, recurrent episode, severe," the appropriate code is:

A) F32.9 - Major depressive disorder, single episode, unspecified B) F33.2 - Major depressive disorder, recurrent severe without psychotic features C) F33.9 - Major depressive disorder, recurrent, unspecified D) F39 - Unspecified mood disorder

Answer: B.

F33.2 accurately captures the recurrent nature and severe degree of the depression as documented.

Reference: ICD-10-CM Tabular List, Category F33

Common Error: Using single episode codes for recurrent depression or missing severity indicators.

Question 117 (Scenario-Based) A patient has "rheumatoid arthritis with rheumatoid heart disease." What is the correct coding approach?

A) M05.30 + I09.9 - Rheumatoid arthritis + Rheumatic heart disease B) M05.30 - Rheumatoid arthritis with rheumatoid heart disease, unspecified site C) M06.9 - Rheumatoid arthritis, unspecified D) I25.10 - Atherosclerotic heart disease

Answer: B.

When rheumatoid arthritis with heart involvement is documented, use the combination code M05.3- rather than separate codes.

Reference: ICD-10-CM Guidelines Section I.C.13.a.3

Common Error: Using separate codes when combination codes are available for systemic diseases with organ involvement.

Question 118 (Multiple Choice) The ICD-10-CM code for "chronic obstructive pulmonary disease" without further specification is:

A) J44.0 - COPD with acute lower respiratory infection B) J44.1 - COPD with acute exacerbation C) J44.9 - Chronic obstructive pulmonary disease, unspecified D) J43.9 - Emphysema, unspecified

Answer: C.

When COPD is documented without additional specification of exacerbation or infection, J44.9 is the appropriate unspecified code.

Reference: ICD-10-CM Tabular List, Category J44

Common Error: Assuming complications that are not documented or using more specific codes without supporting documentation.

Question 119 (Multiple Choice) For risk adjustment purposes, "obesity" should be coded as:

A) E66.9 - Obesity, unspecified B) E66.01 - Morbid obesity due to excess calories C) E66.8 - Other obesity D) Depends on additional documentation

Answer: D.

Obesity coding requires additional information about type (morbid, drug-induced, etc.) and BMI for optimal specificity and HCC mapping.

Reference: ICD-10-CM Guidelines Section I.C.4.a.5

Common Error: Using default obesity codes without seeking additional specificity that may be available.

Question 120 (Select All That Apply) Which coding principles are essential for risk adjustment? (Select all that apply)

A) Code only conditions treated during the encounter B) Code all documented conditions that affect patient care C) Use the most specific code available D) Avoid coding suspected conditions E) Code conditions that require ongoing monitoring

Answer: B, C, D, E.

Risk adjustment coding includes all documented conditions affecting care, uses maximum specificity, avoids suspected conditions, and includes conditions requiring ongoing monitoring. It's not limited to conditions treated in that specific encounter.

Reference: ICD-10-CM Guidelines and CMS Risk Adjustment Guidelines

Common Error: Limiting coding to only conditions actively treated during the encounter.

Question 121 (Multiple Choice) When a patient has both "acute myocardial infarction" and "old myocardial infarction," the coding should be:

A) I21.9 only - Acute myocardial infarction, unspecified B) I25.2 only - Old myocardial infarction C) Both I21.9 and I25.2 D) I22.9 - Subsequent myocardial infarction

Answer: C.

Both conditions should be coded as they represent different clinical situations - the current acute event and the historical condition with ongoing implications.

Reference: ICD-10-CM Guidelines Section I.C.9.e.1

Common Error: Thinking that acute and chronic versions of the same condition are mutually exclusive.

Question 122 (Calculation Exercise) A coder reviews 100 diabetes cases and finds:

- 40 coded as E11.9 (without complications)
- 35 coded as E11.21 (with nephropathy)
- 15 coded as E11.311 (with retinopathy and macular edema)
- 10 coded as E11.40 (with neuropathy)

How many cases have diabetes with complications?

A) 40 B) 60 C) 75 D) 100

Answer: B.

Cases with complications: 35 (nephropathy) + 15 (retinopathy) + 10 (neuropathy) = 60 cases.

Reference: ICD-10-CM Diabetes Coding Guidelines

Common Error: Including uncomplicated diabetes in the complication count.

Question 123 (Multiple Choice) The seventh character in ICD-10-CM codes typically indicates:

A) Severity of the condition B) Laterality (left vs right) C) Encounter type (initial, subsequent, sequela) D) Provider specialty

Answer: C.

The seventh character, when required, typically indicates the type of encounter (initial, subsequent, sequela) particularly for injuries and external causes.

Reference: ICD-10-CM Guidelines Section I.B.2

Common Error: Confusing seventh character meanings with other coding elements like laterality or severity.

Question 124 (Multiple Choice) For "pressure ulcer of sacral region, stage 3," the correct code is:

A) L89.153 - Pressure ulcer of sacral region, stage 3 B) L89.150 - Pressure ulcer of sacral region, unspecified stage C) L89.143 - Pressure ulcer of buttock, stage 3 D) L98.499 - Non-pressure chronic ulcer of skin of other sites with unspecified severity

Answer: A.

L89.153 specifically identifies the sacral location and stage 3 severity of the pressure ulcer.

Reference: ICD-10-CM Tabular List, Category L89

Common Error: Using unspecified stage codes when the stage is clearly documented.

Question 125 (Scenario-Based) A patient has documented "atrial fibrillation, chronic, with rapid ventricular response." What is the most appropriate code?

A) I48.0 - Paroxysmal atrial fibrillation B) I48.1 - Persistent atrial fibrillation C) I48.2 - Chronic atrial fibrillation D) I48.91 - Unspecified atrial fibrillation

Answer: C.

The documentation specifies "chronic" atrial fibrillation, making I48.2 the most appropriate code regardless of the rapid ventricular response.

Reference: ICD-10-CM Tabular List, Category I48

Common Error: Being distracted by additional descriptors (rapid ventricular response) when the primary type (chronic) is clearly stated.

Question 126 (Multiple Choice) The ICD-10-CM guideline for coding "impending" conditions states:

A) Code as if the condition occurred B) Use Z codes for high risk C) Code the underlying condition only D) Do not code impending conditions

Answer: D.

Impending or threatened conditions that do not occur should not be coded. Only code confirmed conditions or symptoms actually present.

Reference: ICD-10-CM Guidelines Section I.C.18.c

Common Error: Coding threatened conditions as if they actually occurred.

Question 127 (Select All That Apply) Which documentation supports coding chronic kidney disease? (Select all that apply)

A) "Decreased kidney function" B) "CKD stage 3" C) "Chronic renal insufficiency" D) "History of acute kidney injury" E) "Elevated creatinine"

Answer: B, C.

Specific documentation of "CKD stage 3" or "chronic renal insufficiency" supports CKD coding. General terms like "decreased function" or "elevated creatinine" without establishing chronicity are insufficient.

Reference: ICD-10-CM Guidelines Section I.C.14.a.2

Common Error: Coding CKD based on lab values or general kidney function descriptions without clear chronic disease documentation.

Question 128 (Multiple Choice) For "Type 2 diabetes with chronic kidney disease, stage 4," the correct coding is:

A) E11.22 - Type 2 diabetes mellitus with diabetic chronic kidney disease B) E11.22 + N18.4 - Type 2 diabetes with CKD + CKD stage 4 C) E11.9 + N18.4 - Type 2 diabetes without complications + CKD stage 4 D) N18.4 only - Chronic kidney disease, stage 4

Answer: B.

When diabetes and CKD coexist, use the diabetes combination code (E11.22) plus the specific CKD stage code (N18.4) to capture both conditions fully.

Reference: ICD-10-CM Guidelines Section I.C.4.a.6.b.ii

Common Error: Using only the combination code without the additional specificity code for CKD stage.

Question 129 (Multiple Choice) The term "acute on chronic" in heart failure coding means:

A) The condition alternates between acute and chronic B) An acute episode is superimposed on a chronic condition C) The condition is both acute and chronic simultaneously D) The condition has changed from acute to chronic

Answer: B.

"Acute on chronic" indicates an acute exacerbation or worsening of an underlying chronic condition.

Reference: ICD-10-CM Guidelines Section I.C.9.a.3

Common Error: Misunderstanding the temporal relationship between acute and chronic elements.

Question 130 (Multiple Choice) When coding "diabetic peripheral neuropathy," the appropriate code is:

A) E11.40 - Type 2 diabetes mellitus with diabetic neuropathy, unspecified B) E11.42 - Type 2 diabetes mellitus with diabetic polyneuropathy C) G63 - Diabetic polyneuropathy D) Either A or B depending on documentation specificity

Answer: D.

If documentation specifies polyneuropathy, use E11.42. If it only states "neuropathy" without further specification, use E11.40.

Reference: ICD-10-CM Guidelines Section I.C.4.a.6.b

Common Error: Always using the most specific code available without considering whether the documentation supports that level of specificity.

Question 131 (Scenario-Based) A patient has "rheumatoid arthritis of multiple joints with rheumatoid factor positive." What is the correct code?

A) M05.79 - Rheumatoid arthritis with rheumatoid factor of unspecified site B) M05.19 - Rheumatoid lung disease with rheumatoid arthritis of unspecified site C) M06.9 - Rheumatoid arthritis, unspecified D) M79.3 - Panniculitis, unspecified

Answer: A.

M05.79 captures rheumatoid arthritis with positive rheumatoid factor. The "multiple joints" documentation supports using "unspecified site" rather than trying to specify each joint.

Reference: ICD-10-CM Tabular List, Category M05

Common Error: Trying to code each individual joint when "multiple joints" or "polyarticular" is documented.

Question 132 (Multiple Choice) The ICD-10-CM code for "morbid obesity" is:

A) E66.01 - Morbid (severe) obesity due to excess calories B) E66.09 - Other obesity due to excess calories C) E66.9 - Obesity, unspecified D) E66.3 - Overweight

Answer: A.

E66.01 specifically represents morbid obesity, which typically indicates BMI ≥40 or BMI ≥35 with comorbidities.

Reference: ICD-10-CM Tabular List, Category E66

Common Error: Using general obesity codes when the documentation specifies morbid obesity.

Question 133 (Multiple Choice) For coding "chronic systolic and diastolic heart failure," the appropriate code is:

A) I50.22 - Chronic systolic heart failure B) I50.32 - Chronic diastolic heart failure C) I50.42 - Chronic combined systolic and diastolic heart failure D) I50.9 - Heart failure, unspecified

Answer: C.

When both systolic and diastolic heart failure are documented as chronic, use the combined code I50.42.

Reference: ICD-10-CM Tabular List, Category I50

Common Error: Coding systolic and diastolic components separately when a combination code exists.

Question 134 (Select All That Apply) Which conditions require additional codes in ICD-10-CM? (Select all that apply)

A) Diabetes with chronic kidney disease B) Pressure ulcer with infection C) Heart failure due to hypertension D) COPD with acute exacerbation E) Rheumatoid arthritis with lung involvement

Answer: A, B.

Diabetes with CKD requires both the diabetes combination code and the specific CKD stage code. Pressure ulcers with infection require both the ulcer code and infection code. The others have complete combination codes available.

Reference: ICD-10-CM Guidelines Multiple Coding Requirements

Common Error: Either over-coding with unnecessary additional codes or under-coding when additional codes are required.

Question 135 (Multiple Choice) The difference between "acute exacerbation" and "acute infection" in COPD is:

A) They are the same condition B) Exacerbation is worsening without infection; infection involves bacterial/viral pathogens C) Exacerbation is less severe than infection D) Infection always requires hospitalization

Answer: B.

Acute exacerbation (J44.1) represents worsening of COPD symptoms without infectious cause, while acute infection (J44.0) involves bacterial or viral respiratory pathogens.

Reference: ICD-10-CM Guidelines Section I.C.10.a.1

Common Error: Using the terms interchangeably when they represent distinct clinical conditions with different codes.

Intermediate Level

Question 136 (Multiple Choice) In CMS-HCC Model V28, which diabetes code provides the highest risk adjustment value?

A) E11.9 - Type 2 diabetes mellitus without complications B) E11.21 - Type 2 diabetes mellitus with diabetic nephropathy C) E11.311 - Type 2 diabetes mellitus with unspecified diabetic retinopathy with macular edema D) E11.40 - Type 2 diabetes mellitus with diabetic neuropathy, unspecified

Answer: B.

Diabetes with nephropathy typically maps to higher-value HCCs than other diabetic complications due to the associated increased medical costs and complexity.

Reference: CMS-HCC Model V28 Coefficient Tables

Common Error: Assuming all diabetic complications have equal risk adjustment value.

Question 137 (Scenario-Based) A patient has documented "Type 2 diabetes with diabetic chronic kidney disease" and separately "chronic kidney disease, stage 4." How should this be coded?

A) E11.22 only B) E11.22 + N18.4 C) E11.9 + N18.4 D) N18.4 only

Answer: B.

Use E11.22 for diabetes with diabetic chronic kidney disease, plus N18.4 to specify the CKD stage, as per ICD-10-CM guidelines requiring both codes.

Reference: ICD-10-CM Guidelines Section I.C.4.a.6.b.ii

Common Error: Using only the combination code without the additional specificity required for CKD staging.

Question 138 (Multiple Choice) Which documentation best supports coding major depressive disorder for risk adjustment?

A) "Patient appears sad" B) "Depression, recurrent, severe" C) "Feeling down lately" D) "Possible depression"

Answer: B.

Clear diagnostic statements with specificity regarding recurrence and severity provide the best support for accurate coding and HCC mapping.

Reference: ICD-10-CM Guidelines Section I.A.19

Common Error: Coding based on symptoms or uncertain diagnoses rather than confirmed conditions.

Question 139 (Select All That Apply) Which ICD-10-CM codes for heart failure map to the same HCC? (Select all that apply)

A) I50.1 - Left ventricular failure, unspecified B) I50.20 - Unspecified systolic heart failure C) I50.30 - Unspecified diastolic heart failure D) I50.810 - Right heart failure, unspecified E) I50.9 - Heart failure, unspecified

Answer: A, B, C, E.

Most heart failure codes map to HCC 221 (Heart failure), with some exceptions. Right heart failure may map differently depending on specific model mapping.

Reference: CMS-HCC Model V28 ICD-10-CM Mapping

Common Error: Assuming all cardiac conditions map to the same HCC without checking specific mappings.

Question 140 (Calculation Exercise) A coding audit reviews 200 chronic kidney disease cases:

- 50 coded as N18.3 (stage 3)
- 75 coded as N18.4 (stage 4)

- 40 coded as N18.5 (stage 5)
- 25 coded as N18.6 (ESRD)
- 10 coded as N19 (unspecified)

What percentage were coded to the highest specificity available?

A) 95% B) 90% C) 85% D) Cannot determine from given information

Answer: D.

Cannot determine highest specificity without knowing what documentation was available to support more specific coding in each case.

Reference: ICD-10-CM Guidelines Section I.A.15

Common Error: Assuming unspecified codes always represent poor coding without considering available documentation.

Question 141 (Multiple Choice) For "acute respiratory failure with hypercapnia," the correct code is:

A) J96.00 - Acute respiratory failure, unspecified whether with hypoxia or hypercapnia B) J96.02 - Acute respiratory failure with hypercapnia C) J96.20 - Acute and chronic respiratory failure, unspecified D) J44.1 - COPD with acute exacerbation

Answer: B.

J96.02 specifically captures acute respiratory failure with hypercapnia, providing the required specificity for the documented condition.

Reference: ICD-10-CM Tabular List, Category J96

Common Error: Using unspecified codes when specific gas exchange abnormalities are documented.

Question 142 (Scenario-Based) A patient with rheumatoid arthritis develops lung nodules documented as "rheumatoid lung disease." What is the appropriate coding?

A) M05.10 - Rheumatoid lung disease with rheumatoid arthritis of unspecified site B) M06.9 + J84.10 - Rheumatoid arthritis + Pulmonary fibrosis C) M05.9 - Rheumatoid arthritis with rheumatoid factor, unspecified D) J94.8 - Other specified pleural conditions

Answer: A.

When rheumatoid arthritis involves the lungs, use the specific combination code M05.1- rather than separate codes for arthritis and lung disease.

Reference: ICD-10-CM Guidelines Section I.C.13.a.3

Common Error: Using separate codes when systemic disease combination codes are available.

Question 143 (Multiple Choice) The coding difference between "chronic kidney disease" and "chronic renal failure" is:

A) They require different ICD-10-CM codes B) CKD includes staging while CRF does not C) They are coded the same way D) CRF is more severe than CKD

Answer: C.

Both terms refer to the same condition and are coded using the N18.- category with appropriate staging when documented.

Reference: ICD-10-CM Guidelines Section I.C.14.a.2

Common Error: Thinking different terminology requires different coding approaches.

Question 144 (Select All That Apply) Which elements are required for complete heart failure coding? (Select all that apply)

A) Acute vs chronic B) Systolic vs diastolic vs combined C) Left vs right sided D) Ejection fraction percentage E) Underlying cause

Answer: A, B.

ICD-10-CM heart failure coding primarily requires acuity (acute vs chronic) and type (systolic vs diastolic vs combined). Other elements may be clinically relevant but aren't required for basic coding.

Reference: ICD-10-CM Tabular List, Category I50

Common Error: Including clinically relevant but not coding-required elements as mandatory for code assignment.

Question 145 (Multiple Choice) For "diabetes type 2 with diabetic gastroparesis," the appropriate code is:

A) E11.43 - Type 2 diabetes mellitus with diabetic autonomic neuropathy B) E11.44 - Type 2 diabetes mellitus with diabetic amyotrophy C) E11.49 - Type 2 diabetes mellitus with other diabetic neurological complication D) E11.8 - Type 2 diabetes mellitus with unspecified complications

Answer: A.

Diabetic gastroparesis is a form of autonomic neuropathy, making E11.43 the most appropriate code.

Reference: ICD-10-CM Tabular List, Category E11

Common Error: Using other neurological complication codes when specific types like autonomic neuropathy are available.

Question 146 (Multiple Choice) The ICD-10-CM code for "hypertensive chronic kidney disease, stage 4" is:

A) I12.9 + N18.4 - Hypertensive CKD without heart failure + CKD stage 4 B) I10 + N18.4 - Essential hypertension + CKD stage 4 C) I12.0 + N18.4 - Hypertensive CKD with heart failure + CKD stage 4 D) N18.4 only - Chronic kidney disease, stage 4

Answer: A.

When hypertension and CKD are documented together, assume causal relationship and use I12.9 (without heart failure) plus the specific CKD stage code.

Reference: ICD-10-CM Guidelines Section I.C.9.a.2

Common Error: Coding hypertension and CKD as separate unrelated conditions.

Question 147 (Scenario-Based) A patient has documented "COPD with emphysema and chronic bronchitis." What is the most appropriate coding approach?

A) J44.9 - COPD, unspecified B) J43.9 + J42 - Emphysema + Chronic bronchitis C) J44.1 - COPD with acute exacerbation D) J44.9 - COPD encompasses both emphysema and chronic bronchitis

Answer: D.

COPD includes both emphysema and chronic bronchitis components, so the COPD code (J44.9) is sufficient without additional component codes.

Reference: ICD-10-CM Guidelines Section I.C.10.a.1

Common Error: Coding individual components when the comprehensive condition code includes all documented elements.

Question 148 (Multiple Choice) For "morbid obesity with BMI 45," the complete coding requires:

A) E66.01 only B) E66.01 + Z68.41 C) Z68.41 only D) E66.01 + E66.9

Answer: B.

Morbid obesity (E66.01) should be accompanied by the specific BMI code (Z68.41 for BMI 40.0-44.9) when documented.

Reference: ICD-10-CM Guidelines Section I.C.4.a.5

Common Error: Omitting BMI codes when specific values are documented.

Question 149 (Multiple Choice) The term "uncontrolled" in diabetes coding means:

A) The patient is non-compliant with treatment B) Blood glucose levels are elevated despite treatment C) The diabetes has progressed to complications D) No specific coding implication

Answer: D.

"Uncontrolled" diabetes does not have specific coding implications in ICD-10-CM and should not be coded to hyperglycemia unless specifically documented as a separate condition.

Reference: ICD-10-CM Guidelines Section I.C.4.a.3

Common Error: Adding hyperglycemia codes based on "uncontrolled" terminology without specific documentation.

Question 150 (Select All That Apply) Which factors affect the specificity of pressure ulcer coding? (Select all that apply)

A) Anatomical location B) Stage or severity C) Healing status D) Underlying cause E) Treatment method

Answer: A, B, C.

Pressure ulcer coding requires location, stage, and healing status. Underlying cause and treatment method don't affect code selection.

Reference: ICD-10-CM Tabular List, Category L89

Common Error: Including treatment or causation factors that don't affect code assignment.

Question 151 (Multiple Choice) For "acute on chronic respiratory failure with hypoxia," the appropriate code is:

A) J96.00 - Acute respiratory failure B) J96.21 - Acute and chronic respiratory failure with hypoxia C) J96.01 + J96.11 - Acute with hypoxia + Chronic with hypoxia D) J96.20 - Acute and chronic respiratory failure, unspecified

Answer: B.

"Acute on chronic" respiratory failure with documented hypoxia should be coded as J96.21.

Reference: ICD-10-CM Tabular List, Category J96

Common Error: Using separate codes for acute and chronic components when combination codes exist.

Question 152 (Scenario-Based) A patient with diabetes develops a foot ulcer documented as "diabetic foot ulcer, right great toe." What coding is required?

A) E11.621 only - Type 2 diabetes with foot ulcer B) E11.621 + L97.511 - Diabetes with foot ulcer + Non-pressure ulcer of right great toe with fat layer exposed C) L97.511 only - Non-pressure ulcer code D) Cannot code without more specific ulcer staging

Answer: B.

Diabetic foot ulcers require both the diabetes combination code (E11.621) and the specific ulcer code with location and severity when documented.

Reference: ICD-10-CM Guidelines Section I.C.4.a.6.b.ii

Common Error: Using only the diabetes combination code without the required additional ulcer specificity code.

Question 153 (Multiple Choice) The coding for "bipolar disorder, current episode manic severe with psychotic features" is:

A) F31.2 - Bipolar disorder, current episode manic severe without psychotic features B) F31.5 - Bipolar disorder, current episode depressed, severe, with psychotic features C) F31.64 - Bipolar disorder, current episode mixed, severe, with psychotic features D) F31.13 - Bipolar disorder, current episode manic severe with psychotic features

Answer: D.

F31.13 specifically captures the current manic episode that is severe with psychotic features.

Reference: ICD-10-CM Tabular List, Category F31

Common Error: Misidentifying the current episode type or psychotic feature presence.

Question 154 (Multiple Choice) For "chronic systolic heart failure due to ischemic cardiomyopathy," the coding should be:

A) I50.22 only - Chronic systolic heart failure B) I25.5 only - Ischemic cardiomyopathy C) I25.5 + I50.22 - Both conditions D) I50.1 - Left ventricular failure

Answer: C.

Both the underlying ischemic cardiomyopathy (I25.5) and the resulting chronic systolic heart failure (I50.22) should be coded.

Reference: ICD-10-CM Guidelines Section I.C.9.a.3

Common Error: Coding only the manifestation (heart failure) without the underlying cause (ischemic cardiomyopathy).

Question 155 (Calculation Exercise) A plan's coding review finds the following diabetes coding patterns:

- 300 cases coded as diabetes without complications
- 150 cases coded as diabetes with nephropathy
- 100 cases coded as diabetes with retinopathy
- 50 cases coded as diabetes with neuropathy

What percentage of diabetes cases have documented complications?

A) 25% B) 50% C) 75% D) 100%

Answer: B.

Complicated cases: 150 + 100 + 50 = 300. Total cases: 300 + 300 = 600. Percentage: 300/600 = 50%.

Reference: Diabetes Complication Coding Analysis

Common Error: Including uncomplicated cases in the complication percentage.

Question 156 (Multiple Choice) The ICD-10-CM code for "chronic diastolic heart failure" is:

A) I50.30 - Unspecified diastolic heart failure B) I50.31 - Acute diastolic heart failure C) I50.32 - Chronic diastolic heart failure D) I50.33 - Acute on chronic diastolic heart failure

Answer: C.

I50.32 specifically represents chronic diastolic heart failure as documented.

Reference: ICD-10-CM Tabular List, Category I50

Common Error: Using unspecified codes when chronicity is clearly documented.

Question 157 (Select All That Apply) Which documentation elements support coding chronic kidney disease stage 4? (Select all that apply)

A) "CKD stage 4" B) "Severe chronic kidney disease" C) "GFR 25" D) "Chronic renal insufficiency, severe" E) "Pre-dialysis kidney disease"

Answer: A, B, D.

Explicit stage documentation, severity descriptors, and clinical terms like "severe chronic renal insufficiency" support stage 4 coding. GFR values alone or pre-dialysis terms require physician interpretation.

Reference: ICD-10-CM Guidelines Section I.C.14.a.2

Common Error: Coding CKD stages based solely on lab values without physician documentation of the stage.

Question 158 (Multiple Choice) For "rheumatoid arthritis with rheumatoid heart disease," the appropriate code is:

A) M05.30 - Rheumatoid arthritis with rheumatoid heart disease, unspecified site B) M06.9 + I09.9 - Rheumatoid arthritis + Rheumatic heart disease C) M05.9 - Rheumatoid arthritis with rheumatoid factor D) I25.10 - Atherosclerotic heart disease

Answer: A.

Use the combination code M05.30 for rheumatoid arthritis with cardiac involvement rather than separate codes.

Reference: ICD-10-CM Tabular List, Category M05

Common Error: Using separate codes when combination codes for systemic disease with organ involvement are available.

Question 159 (Scenario-Based) A patient has "Type 1 diabetes with diabetic chronic kidney disease" and "chronic kidney disease, stage 3." How should this be coded?

A) E10.22 + N18.3 B) E10.9 + N18.3 C) E10.22 only D) N18.3 only

Answer: A.

Use the Type 1 diabetes with diabetic CKD code (E10.22) plus the specific CKD stage code (N18.3) for complete coding.

Reference: ICD-10-CM Guidelines Section I.C.4.a.6.b.ii

Common Error: Forgetting to use Type 1 diabetes codes when Type 1 is specifically documented.

Question 160 (Multiple Choice) The coding for "major depressive disorder, single episode, severe with psychotic features" is:

A) F32.3 - Major depressive disorder, single episode, severe with psychotic symptoms B) F32.2 - Major depressive disorder, single episode, severe without psychotic features C) F33.3 - Major depressive disorder, recurrent, severe with psychotic symptoms D) F32.9 - Major depressive disorder, single episode, unspecified

Answer: A.

F32.3 specifically captures single episode severe depression with psychotic features.

Reference: ICD-10-CM Tabular List, Category F32

Common Error: Confusing single episode vs recurrent or missing psychotic feature documentation.

Question 161 (Multiple Choice) For "COPD with acute respiratory failure," the coding should be:

A) J44.1 only - COPD with acute exacerbation B) J44.0 + J96.00 - COPD with infection + Acute respiratory failure C) J44.9 + J96.00 - COPD + Acute respiratory failure D) J96.00 only - Acute respiratory failure

Answer: C.

Both COPD (J44.9) and acute respiratory failure (J96.00) should be coded as they represent different conditions that can coexist.

Reference: ICD-10-CM Guidelines Section I.C.10.a.1

Common Error: Assuming COPD codes include all respiratory complications.

Question 162 (Multiple Choice) The term "diabetic" as a modifier in "diabetic nephropathy" indicates:

A) The patient has diabetes and separately has nephropathy B) The nephropathy is caused by diabetes C) The conditions occur simultaneously D) No specific relationship

Answer: B.

"Diabetic" as a modifier indicates causal relationship, requiring use of diabetes combination codes rather than separate condition codes.

Reference: ICD-10-CM Guidelines Section I.C.4.a.6.b

Common Error: Treating diabetic complications as separate unrelated conditions.

Question 163 (Select All That Apply) Which conditions commonly suppress other conditions in HCC hierarchies? (Select all that apply)

A) Diabetes with complications B) Acute myocardial infarction C) Chronic kidney disease stage 5 D) Morbid obesity E) Major depressive disorder

Answer: A, C.

More severe forms of diabetes and advanced CKD typically suppress less severe forms in hierarchical relationships. Acute MI, morbid obesity, and major depression don't typically have hierarchical suppressions.

Reference: CMS-HCC Model V28 Hierarchy Tables

Common Error: Assuming all severe conditions suppress related less severe conditions without checking specific hierarchy relationships.

Question 164 (Multiple Choice) For "pressure ulcer of left heel, stage 2, healing," the correct code is:

A) L89.622 - Pressure ulcer of left heel, stage 2 B) L89.622 - Pressure ulcer of left heel, stage 2 (healing status not coded separately) C) L89.620 - Pressure ulcer of left heel, unspecified stage D) L89.602 - Pressure ulcer of left heel, stage 2, healing

Answer: A.

Stage 2 pressure ulcer of left heel is coded as L89.622. Healing status is not separately coded in current ICD-10-CM.

Reference: ICD-10-CM Tabular List, Category L89

Common Error: Trying to code healing status separately when it's not part of the ICD-10-CM classification system.

Question 165 (Scenario-Based) A patient has documented "congestive heart failure, acute, due to hypertensive heart disease." What is the appropriate coding?

A) I11.0 + I50.9 - Hypertensive heart disease with heart failure + Heart failure B) I11.0 only - Hypertensive heart disease with heart failure C) I50.1 - Left ventricular failure D) I10 + I50.9 - Essential hypertension + Heart failure

Answer: B.

I11.0 (Hypertensive heart disease with heart failure) includes the heart failure, so no additional heart failure code is needed.

Reference: ICD-10-CM Guidelines Section I.C.9.a.1

Common Error: Adding additional heart failure codes when the hypertensive heart disease code already includes heart failure.

Question 166 (Multiple Choice) The coding for "acute respiratory failure with hypoxia and hypercapnia" is:

A) J96.01 - Acute respiratory failure with hypoxia B) J96.02 - Acute respiratory failure with hypercapnia C) J96.00 - Acute respiratory failure, unspecified whether with hypoxia or hypercapnia D) J96.01 + J96.02 - Both conditions coded separately

Answer: C.

When both hypoxia and hypercapnia are present, use the unspecified code J96.00 as there is no combination code for both conditions.

Reference: ICD-10-CM Tabular List, Category J96

Common Error: Trying to code both gas exchange abnormalities separately when a single code is appropriate.

Question 167 (Multiple Choice) For "chronic combined systolic and diastolic heart failure," the appropriate code is:

A) I50.22 + I50.32 - Chronic systolic + Chronic diastolic B) I50.42 - Chronic combined systolic and diastolic heart failure C) I50.9 - Heart failure, unspecified D) I50.20 + I50.30 - Unspecified systolic + Unspecified diastolic

Answer: B.

When both systolic and diastolic heart failure are chronic, use the specific combination code I50.42.

Reference: ICD-10-CM Tabular List, Category I50

Common Error: Using separate codes when specific combination codes are available.

Question 168 (Calculation Exercise) A coding quality review examines 150 heart failure cases:

- 60 coded as I50.9 (unspecified)
- 40 coded as I50.22 (chronic systolic)
- 30 coded as I50.32 (chronic diastolic)
- 20 coded as I50.42 (chronic combined)

What is the specificity rate (non-unspecified codes)?

A) 40% B) 60% C) 90% D) 20%

Answer: B.

Specific codes: 40 + 30 + 20 = 90. Specificity rate: 90/150 = 60%.

Reference: Coding Quality Metrics

Common Error: Calculating based on total codes rather than specific vs unspecified ratios.

Question 169 (Multiple Choice) The ICD-10-CM code for "end stage renal disease on dialysis" is:

A) N18.6 only - End stage renal disease B) N18.6 + Z99.2 - ESRD + Dependence on renal dialysis C) Z99.2 only - Dependence on renal dialysis D) N18.5 - Chronic kidney disease, stage 5

Answer: B.

ESRD (N18.6) should be coded along with dialysis status (Z99.2) when both are documented to provide complete clinical picture.

Reference: ICD-10-CM Guidelines Section I.C.14.a.2

Common Error: Using only the ESRD code without capturing the dialysis dependence when documented.

Question 170 (Multiple Choice) For "diabetic polyneuropathy, type 2," the appropriate code is:

A) E11.40 - Type 2 diabetes with diabetic neuropathy, unspecified B) E11.42 - Type 2 diabetes with diabetic polyneuropathy C) G63 - Diabetic polyneuropathy D) E11.8 - Type 2 diabetes with unspecified complications

Answer: B.

When polyneuropathy is specifically documented, use E11.42 rather than the unspecified neuropathy code.

Reference: ICD-10-CM Tabular List, Category E11

Common Error: Using unspecified complication codes when specific types are documented and have specific codes available.

Question 171 (Select All That Apply) Which elements are essential for accurate heart failure coding? (Select all that apply)

A) Ejection fraction percentage B) Acute vs chronic designation C) Systolic vs diastolic type D) Underlying etiology E) New York Heart Association class

Answer: B, C.

Acuity (acute vs chronic) and type (systolic vs diastolic) are essential for ICD-10-CM heart failure coding. Other elements are clinically relevant but not required for basic coding.

Reference: ICD-10-CM Guidelines Section I.C.9.a.3

Common Error: Including clinically relevant but not coding-essential elements as requirements.

Question 172 (Multiple Choice) The coding for "acute on chronic kidney disease, stage 4" is:

A) N17.9 + N18.4 - Acute kidney failure + CKD stage 4 B) N18.4 only - Chronic kidney disease, stage 4 C) N19 - Unspecified kidney failure D) N17.9 only - Acute kidney failure

Answer: B.

"Acute on chronic kidney disease" refers to acute worsening of chronic disease, not separate acute and chronic conditions. Code the chronic condition with appropriate staging.

Reference: ICD-10-CM Guidelines Section I.C.14.a.1

Common Error: Coding acute and chronic kidney conditions separately when documentation indicates acute exacerbation of chronic disease.

Question 173 (Scenario-Based) A patient with diabetes develops peripheral arterial disease documented as "diabetic peripheral angiopathy." What is the correct coding?

A) E11.51 - Type 2 diabetes with diabetic peripheral angiopathy without gangrene B) E11.9 + I73.9 - Diabetes without complications + Peripheral arterial disease C) I73.9 only - Peripheral arterial disease D) E11.8 - Type 2 diabetes with unspecified complications

Answer: A.

Diabetic peripheral angiopathy should be coded with the specific diabetes combination code E11.51.

Reference: ICD-10-CM Guidelines Section I.C.4.a.6.b

Common Error: Coding diabetic vascular complications as separate unrelated conditions.

Question 174 (Multiple Choice) For "manic episode with psychotic features, severe," in a patient with no prior mood episodes, the code is:

A) F30.2 - Manic episode, severe without psychotic symptoms B) F30.8 - Other manic episodes C) F30.2 - Manic episode, severe with psychotic symptoms D) F31.13 - Bipolar disorder, current episode manic severe with psychotic symptoms

Answer: C.

A single manic episode with psychotic features is coded as F30.2. Bipolar disorder requires evidence of prior mood episodes.

Reference: ICD-10-CM Tabular List, Category F30

Common Error: Assuming any manic episode indicates bipolar disorder without considering episode history.

Question 175 (Multiple Choice) The most specific code for "chronic obstructive pulmonary disease with emphysema" is:

A) J43.9 - Emphysema, unspecified B) J44.9 - Chronic obstructive pulmonary disease, unspecified C) J44.0 - COPD with acute lower respiratory infection D) J44.1 - COPD with acute exacerbation

Answer: B.

COPD encompasses emphysema, so the COPD code is more comprehensive and appropriate than coding emphysema separately.

Reference: ICD-10-CM Guidelines Section I.C.10.a.1

Common Error: Coding components of COPD separately when the comprehensive COPD diagnosis includes those components.

Advanced Level

Question 176 (Multiple Choice) In the transition from ICD-10-CM V24 to V28 HCC model, which coding change had the most significant impact on diabetes risk adjustment?

A) Consolidation of diabetes HCC categories B) Elimination of diabetes interaction terms C) Changes in diabetes hierarchy relationships D) Introduction of diabetes severity modifiers

Answer: A.

V28 consolidated multiple diabetes HCCs into fewer categories, significantly changing risk adjustment values for various diabetes manifestations.

Reference: CMS-HCC Model V28 Diabetes Category Changes

Common Error: Focusing on interaction or hierarchy changes rather than the fundamental category consolidation.

Question 177 (Scenario-Based) A Medicare Advantage plan's advanced analytics identifies a provider who codes diabetes complications in 85% of diabetic patients, while the network average is 35%. The provider's documentation consistently supports the coding. What is the most appropriate interpretation?

A) The provider is likely overcoding inappropriately B) The provider serves a sicker diabetic population or has superior clinical assessment C) The coding should be adjusted to match network averages D) An immediate audit is warranted

Answer: B.

Higher complication rates with supporting documentation may indicate either case mix differences or superior clinical assessment and documentation practices.

Reference: Risk Adjustment Variation Analysis Best Practices

Common Error: Automatically assuming variation indicates inappropriate activity without considering legitimate clinical or practice differences.

Question 178 (Multiple Choice) The most sophisticated approach to ICD-10-CM coding optimization for risk adjustment involves:

A) Focusing only on highest-value HCC codes B) Integrating clinical decision support with real-time HCC mapping C) Training coders to memorize HCC mappings D) Implementing automated coding software

Answer: B.

Advanced optimization integrates clinical decision support that shows real-time HCC implications to support both clinical care and appropriate documentation.

Reference: Advanced Health Information Technology in Risk Adjustment

Common Error: Focusing on memorization or automation rather than intelligent decision support integration.

Question 179 (Calculation Exercise) A plan implements AI-powered coding assistance that increases HCC capture rate from 65% to 78% across 5,000 members with chronic conditions. If the average HCC value is $2,500 annually, what is the financial impact?

A) $812,500 B) $325,000 C) $1,625,000 D) $487,500

Answer: C.

Improvement: 78% - 65% = 13% increase. Additional capture: 5,000 × 13% × $2,500 = $1,625,000.

Reference: AI Implementation ROI Analysis

Common Error: Calculating based on total population rather than the incremental improvement percentage.

Question 180 (Select All That Apply) Advanced ICD-10-CM coding for risk adjustment should consider: (Select all that apply)

A) Temporal patterns of diagnosis documentation B) Provider specialty-specific coding patterns C) Integration with clinical quality measures D) Predictive modeling for diagnosis progression E) Geographic variations in disease prevalence

Answer: A, B, C, D.

Advanced coding considers temporal patterns, specialty variations, quality integration, and predictive progression. Geographic prevalence affects population health but not individual coding decisions.

Reference: Advanced Risk Adjustment Coding Analytics

Common Error: Including population-level factors that don't affect individual patient coding decisions.

Question 181 (Multiple Choice) Machine learning applications in ICD-10-CM coding are most effective for:

A) Replacing human coders entirely B) Identifying potential coding gaps and opportunities C) Automatically assigning codes without documentation D) Eliminating the need for physician documentation

Answer: B.

ML excels at pattern recognition to identify potential coding opportunities and gaps that humans might miss due to data volume and complexity.

Reference: Machine Learning in Healthcare Coding

Common Error: Expecting ML to replace rather than augment human expertise and documentation requirements.

Question 182 (Scenario-Based) A health plan's predictive model identifies that members with diabetes + depression have 30% higher costs than diabetes alone, but only 40% of diabetic members are screened for depression. What is the optimal coding strategy?

A) Code depression for all diabetic members B) Implement systematic depression screening for diabetic members C) Focus only on members already diagnosed with depression D) Ignore the depression relationship

Answer: B.

Systematic screening leads to appropriate clinical identification and documentation of comorbid conditions that affect both care needs and risk adjustment.

Reference: Evidence-Based Comorbidity Screening Programs

Common Error: Focusing on coding outcomes rather than appropriate clinical screening that leads to accurate coding.

Question 183 (Multiple Choice) The most advanced ICD-10-CM coding quality metrics include:

A) Volume of codes assigned per encounter B) Specificity rates adjusted for available documentation C) Speed of code assignment D) Consistency with prior year coding

Answer: B.

Advanced quality metrics assess whether coders achieve maximum appropriate specificity given the available documentation, not just overall specificity rates.

Reference: Advanced Coding Quality Assessment

Common Error: Using simple volume or speed metrics rather than documentation-adjusted specificity measures.

Question 184 (Multiple Choice) Natural Language Processing (NLP) applications in risk adjustment coding should prioritize:

A) Extracting all possible diagnosis mentions B) Identifying clinically relevant diagnoses that meet MEAT criteria C) Generating codes automatically without human review D) Replacing physician documentation review

Answer: B.

NLP should identify clinically relevant diagnoses that meet risk adjustment criteria (MEAT) rather than simply extracting all diagnosis mentions.

Reference: NLP Applications in Clinical Documentation

Common Error: Focusing on comprehensive extraction rather than clinically and regulatory relevant identification.

Question 185 (Scenario-Based) A plan's advanced analytics reveal that cardiologists document heart failure in 90% of patients with reduced ejection fraction, while primary care physicians document it in only 60% of similar patients. What intervention would be most effective?

A) Require all cardiology coding review B) Implement clinical decision support for PCPs linking EF results to heart failure consideration C) Transfer all heart failure patients to cardiologists D) Manually add heart failure codes for low EF patients

Answer: B.

Clinical decision support helps PCPs recognize the clinical relationship between objective findings (low EF) and appropriate diagnoses (heart failure).

Reference: Clinical Decision Support for Diagnostic Recognition

Common Error: Choosing administrative solutions rather than clinical education and decision support tools.

Question 186 (Multiple Choice) Sophisticated risk adjustment coding programs integrate:

A) Only diagnostic coding accuracy B) Clinical documentation improvement, coding accuracy, and care management C) Automated coding with minimal human oversight D) Focus solely on highest-value HCCs

Answer: B.

Advanced programs integrate clinical documentation improvement, accurate coding, and care management to optimize both clinical and risk adjustment outcomes.

Reference: Integrated Risk Adjustment Program Design

Common Error: Focusing on single components rather than comprehensive integrated approaches.

Question 187 (Calculation Exercise) A plan's NLP system reviews 10,000 clinical notes and identifies 2,500 potential additional HCC-eligible diagnoses. Manual review confirms 1,875 are appropriate for coding. If 80% of these translate to new HCCs worth an average of $1,800, what is the value?

A) $2,700,000 B) $3,375,000 C) $4,500,000 D) $5,400,000

Answer: A.

Confirmed diagnoses: 1,875. New HCCs: 1,875 × 80% = 1,500. Value: 1,500 × $1,800 = $2,700,000.

Reference: NLP Implementation ROI Analysis

Common Error: Not accounting for the manual review confirmation step or HCC conversion rate.

Question 188 (Select All That Apply) Advanced coding analytics should monitor: (Select all that apply)

A) Code-specific denial rates in RADV audits B) Provider-specific coding pattern variations C) Temporal trends in diagnostic specificity D) Correlation between coded conditions and clinical outcomes E) Geographic coding pattern variations

Answer: A, B, C, D.

Advanced analytics monitor audit performance, provider variations, temporal trends, and clinical correlations. Geographic patterns are less relevant for individual coding decisions.

Reference: Comprehensive Coding Analytics Framework

Common Error: Including geographic factors that affect population health but not individual coding quality.

Question 189 (Multiple Choice) The most sophisticated approach to managing ICD-10-CM hierarchy relationships involves:

A) Memorizing all hierarchy tables B) Implementing real-time hierarchy checking in coding systems C) Avoiding documentation of less severe conditions D) Coding only the most severe condition in each hierarchy

Answer: B.

Real-time hierarchy checking ensures coders understand suppression relationships and focus on clinically appropriate documentation regardless of hierarchy implications.

Reference: Advanced Coding System Design

Common Error: Avoiding appropriate clinical documentation due to hierarchy concerns rather than implementing systematic checking.

Question 190 (Scenario-Based) A health plan's machine learning model predicts which members are likely to develop diabetic complications within the next 6 months with 85% accuracy. How should this inform coding strategy?

A) Pre-code complications for high-risk members B) Focus clinical screenings and follow-up on high-risk members C) Adjust risk scores prospectively based on predictions D) Use predictions to validate existing coding

Answer: B.

Predictive models should inform clinical interventions and screening strategies that may lead to earlier appropriate identification and documentation of developing conditions.

Reference: Predictive Analytics in Clinical Care

Common Error: Using predictions to directly influence coding rather than clinical care that leads to appropriate documentation.

Question 191 (Multiple Choice) Advanced risk adjustment coding compliance includes:

A) Automated flagging of statistically unusual coding patterns B) Manual review of every coded diagnosis C) Restriction to only commonly coded conditions D) Elimination of complex combination codes

Answer: A.

Advanced compliance uses statistical analysis to flag unusual patterns that warrant investigation while maintaining efficiency through targeted rather than universal manual review.

Reference: Advanced Compliance Monitoring Systems

Common Error: Relying on manual processes that don't scale effectively or restricting appropriate coding complexity.

Question 192 (Multiple Choice) The future of ICD-10-CM coding in risk adjustment will likely emphasize:

A) Simplified coding with fewer options B) Integration with real-time clinical data and AI assistance C) Elimination of physician documentation requirements D) Focus only on inpatient diagnoses

Answer: B.

Future developments will likely integrate real-time clinical data with AI assistance to support more accurate and timely coding while maintaining documentation requirements.

Reference: Healthcare Information Technology Evolution

Common Error: Expecting simplification rather than increased sophistication with better technological support.

Question 193 (Scenario-Based) A plan's advanced analytics identify that members coded with diabetes + CKD + anemia have 40% higher costs than the sum of individual conditions would predict, but this three-way interaction isn't in the HCC model. What is the optimal response?

A) Request CMS add the interaction to the model B) Develop targeted care management for this combination C) Code these conditions as aggressively as possible D) Focus on documenting only two of the three conditions

Answer: B.

Plans should develop care management strategies for clinically meaningful high-risk combinations even when not explicitly recognized in payment models.

Reference: Population Health Management for Complex Comorbidities

Common Error: Waiting for model recognition rather than proactively managing identified high-risk populations.

Question 194 (Multiple Choice) Sophisticated diagnostic coding validation should include:

A) Clinical reasonableness assessment using AI algorithms B) Simple code frequency analysis C) Comparison to national coding averages only D) Focus on highest-volume codes only

Answer: A.

Advanced validation uses AI to assess clinical reasonableness of coded diagnoses given documented clinical evidence, not just statistical comparisons.

Reference: AI-Powered Clinical Validation Systems

Common Error: Relying on statistical comparisons rather than clinical appropriateness assessment.

Question 195 (Calculation Exercise) A plan implements advanced coding technology that improves diagnostic specificity from 75% to 89% across 8,000 chronic condition encounters. If improved specificity increases average HCC value by $150 per encounter, what is the annual impact?

A) $168,000 B) $1,120,000 C) $1,680,000 D) $200,000

Answer: A.

Improvement: 89% - 75% = 14% increase. Impact: 8,000 × 14% × $150 = $168,000.

Reference: Coding Specificity Improvement Analysis

Common Error: Applying the improvement rate to all encounters rather than calculating the incremental improvement.

Question 196 (Select All That Apply) Next-generation ICD-10-CM coding systems should integrate: (Select all that apply)

A) Real-time clinical decision support B) Automated MEAT criteria validation C) Predictive modeling for diagnosis progression D) Social determinants of health data E) Blockchain for code verification

Answer: A, B, C.

Next-generation systems should integrate clinical decision support, MEAT validation, and predictive modeling. SDOH data is valuable for care but doesn't affect coding decisions, and blockchain isn't necessary for coding verification.

Reference: Next-Generation Health Information Systems

Common Error: Including technologies that don't directly enhance coding accuracy or appropriateness.

Question 197 (Multiple Choice) The most advanced approach to coding education for risk adjustment involves:

A) Memorization of code definitions B) Case-based learning with real-time feedback on HCC implications C) Focus on highest-value codes only D) Separation of clinical and coding education

Answer: B.

Advanced education uses case-based learning that shows both clinical reasoning and risk adjustment implications, providing real-time feedback on coding decisions.

Reference: Advanced Health Information Management Education

Common Error: Separating clinical and coding education rather than integrating them for comprehensive understanding.

Question 198 (Scenario-Based) A health plan's AI system identifies potential coding opportunities but flags them for human review due to clinical complexity. The human reviewer confirms 70% of flagged cases are appropriate for additional coding. What does this indicate?

A) The AI system is poorly calibrated B) The AI system is appropriately conservative and effective when combined with human oversight C) Human reviewers are too lenient D) The flagging threshold should be lowered

Answer: B.

A 70% confirmation rate indicates appropriate AI conservatism with effective human-AI collaboration for complex cases requiring clinical judgment.

Reference: Human-AI Collaboration in Healthcare

Common Error: Expecting near-perfect AI accuracy rather than appreciating effective human-AI collaboration.

Question 199 (Multiple Choice) Advanced risk adjustment coding quality measurement should focus on:

A) Code assignment speed B) Clinical appropriateness and documentation support C) Maximization of HCC values D) Consistency with prior years

Answer: B.

Quality measurement should prioritize clinical appropriateness and adequate documentation support rather than speed, maximization, or historical consistency.

Reference: Quality Measurement in Health Information Management

Common Error: Focusing on efficiency or financial metrics rather than clinical appropriateness and accuracy.

Question 200 (Multiple Choice) The integration of social determinants of health (SDOH) with ICD-10-CM coding for risk adjustment should:

A) Replace existing clinical codes B) Supplement clinical coding to provide comprehensive risk assessment C) Focus only on housing and income factors D) Only apply to specific demographic groups

Answer: B.

SDOH integration should supplement clinical coding to provide more comprehensive risk assessment without replacing established clinical coding requirements.

Reference: Social Determinants Integration in Healthcare

Common Error: Viewing SDOH as replacement for rather than enhancement of clinical risk factors.

Documentation & Compliance

Question 201 (Multiple Choice) What does the MEAT acronym stand for in risk adjustment documentation?

A) Medical history, Evaluation, Assessment, Treatment B) Monitoring, Evaluating, Assessing, Treating C) Medical necessity, Evidence, Analysis, Therapy D) Monitoring, Evidence, Assessment, Therapy

Answer: B.

MEAT stands for Monitoring, Evaluating, Assessing, or Treating. At least one of these elements must be documented for each diagnosis to meet risk adjustment requirements.

Reference: CMS Risk Adjustment Data Technical Assistance

Common Error: Confusing MEAT with other documentation frameworks or requiring all four elements instead of at least one.

Question 202 (Multiple Choice) Which provider type CAN document diagnoses for risk adjustment purposes?

A) Registered nurses B) Medical assistants C) Physician assistants D) Physical therapists

Answer: C.

Physician assistants are among the acceptable provider types who can document diagnoses for risk adjustment, along with physicians and nurse practitioners.

Reference: CMS Acceptable Provider Types List

Common Error: Including all healthcare workers rather than specific credentialed provider types authorized for risk adjustment.

Question 203 (Select All That Apply) Which elements are required for a face-to-face encounter in risk adjustment? (Select all that apply)

A) Direct patient contact B) Acceptable provider type C) Documentation of MEAT criteria D) Minimum 15-minute encounter duration E) Hospital setting requirement

Answer: A, B, C.

Face-to-face encounters require direct patient contact, an acceptable provider type, and MEAT criteria documentation. There are no minimum duration or setting requirements.

Reference: CMS Face-to-Face Encounter Requirements

Common Error: Adding duration or setting requirements that don't exist in the regulations.

Question 204 (Multiple Choice) How often must a chronic condition be documented to continue generating HCC payments?

A) Once per lifetime B) Every six months C) Annually D) Only when symptoms are present

Answer: C.

Chronic conditions must be documented at least once per calendar year during a qualifying face-to-face encounter to continue generating risk adjustment payments.

Reference: CMS Annual Documentation Requirements

Common Error: Assuming once documented, conditions continue to generate payments indefinitely.

Question 205 (Scenario-Based) A patient with diabetes visits their primary care physician. The physician documents "diabetes - stable, continue metformin." Does this meet MEAT criteria?

A) Yes - Treatment is documented B) No - Not enough detail provided C) Yes - Monitoring is implied D) No - Assessment is missing

Answer: A.

"Continue metformin" represents Treatment, which satisfies MEAT criteria. Only one of the four MEAT elements is required.

Reference: CMS MEAT Documentation Examples

Common Error: Requiring multiple MEAT elements or more detailed documentation than necessary.

Question 206 (Multiple Choice) Which documentation would NOT support risk adjustment for heart failure?

A) "CHF - doing well on current medications" B) "Heart failure - will increase Lasix dose" C) "Patient reports history of CHF but no current symptoms" D) "Chronic heart failure - monitoring fluid status"

Answer: C.

Historical mention without current management, monitoring, evaluation, or treatment does not meet MEAT criteria for ongoing HCC eligibility.

Reference: CMS Historical vs. Active Condition Guidelines

Common Error: Assuming all historical mentions of conditions qualify for risk adjustment.

Question 207 (Multiple Choice) Telehealth encounters for risk adjustment documentation:

A) Are never acceptable B) Are acceptable if they meet face-to-face requirements C) Only count for certain specialties D) Require additional documentation elements

Answer: B.

Telehealth encounters can qualify for risk adjustment if they constitute face-to-face encounters with acceptable providers and meet MEAT criteria.

Reference: CMS Telehealth Guidelines for Risk Adjustment

Common Error: Excluding telehealth encounters that meet standard face-to-face requirements.

Question 208 (Select All That Apply) Which are examples of "Monitoring" under MEAT criteria? (Select all that apply)

A) Checking blood pressure for hypertension B) Reviewing lab results for diabetes C) Asking about medication side effects D) Scheduling follow-up appointments E) Weighing patient with heart failure

Answer: A, B, C, E.

Monitoring includes checking vital signs, reviewing test results, assessing symptoms, and other clinical observations. Simply scheduling appointments without clinical monitoring doesn't count.

Reference: CMS MEAT Criteria Examples

Common Error: Including administrative activities rather than clinical monitoring activities.

Question 209 (Multiple Choice) The minimum documentation standard for risk adjustment requires:

A) Detailed treatment plans for all conditions B) At least one MEAT element per documented condition C) Complete problem lists in every encounter D) Specialist confirmation of all diagnoses

Answer: B.

The minimum standard requires at least one MEAT element (Monitoring, Evaluating, Assessing, or Treating) for each condition claimed for risk adjustment.

Reference: CMS Minimum Documentation Standards

Common Error: Setting unnecessarily high documentation standards beyond regulatory requirements.

Question 210 (Multiple Choice) Which statement about "Assessment" in MEAT criteria is correct?

A) It must include a formal assessment score B) It requires changing the treatment plan C) It can include clinical decision-making about the condition D) It only applies to mental health conditions

Answer: C.

Assessment includes clinical thinking, decision-making, or evaluation of the condition's status, not necessarily formal scoring or treatment changes.

Reference: CMS MEAT Assessment Examples

Common Error: Limiting assessment to formal tools or requiring treatment modifications.

Question 211 (Scenario-Based) A nurse practitioner documents "COPD - patient using rescue inhaler more frequently, will add long-acting bronchodilator." Which MEAT elements are present?

A) Monitoring only B) Evaluating and Treating C) All four elements D) Assessing and Treating

Answer: B.

"Using rescue inhaler more frequently" represents Evaluating the condition status, and "will add long-acting bronchodilator" represents Treating.

Reference: CMS MEAT Documentation Analysis

Common Error: Missing the evaluation component in symptom assessment or over-identifying MEAT elements.

Question 212 (Multiple Choice) Risk adjustment documentation must occur during:

A) Any healthcare encounter B) Face-to-face encounters with acceptable providers C) Inpatient admissions only D) Annual wellness visits only

Answer: B.

Documentation must occur during face-to-face encounters with acceptable provider types. It's not limited to specific encounter types like wellness visits or inpatient stays.

Reference: CMS Encounter Type Requirements

Common Error: Limiting acceptable encounters to specific types rather than any qualifying face-to-face encounter.

Question 213 (Multiple Choice) When can a condition documented in a previous year continue to generate HCC payments?

A) Always, once documented B) Never, must be documented annually C) Only if it's a chronic condition D) Only if documented in current year

Answer: D.

All conditions, whether acute or chronic, must be documented in the current year to generate current year HCC payments.

Reference: CMS Annual Documentation Requirements

Common Error: Assuming chronic conditions continue to generate payments without current year documentation.

Question 214 (Select All That Apply) Which providers can document conditions for risk adjustment? (Select all that apply)

A) Physicians (MD/DO) B) Nurse practitioners C) Physician assistants D) Registered nurses E) Licensed clinical social workers

Answer: A, B, C.

Physicians, nurse practitioners, and physician assistants are acceptable provider types. Registered nurses and LCSWs cannot document conditions for risk adjustment purposes.

Reference: CMS Acceptable Provider Types

Common Error: Including all licensed healthcare professionals rather than specific provider types authorized for risk adjustment.

Question 215 (Multiple Choice) The phrase "continue current medications" in documentation represents which MEAT element?

A) Monitoring B) Evaluating C) Assessing D) Treating

Answer: D.

Continuing medications represents ongoing Treatment of the condition, satisfying the "T" in MEAT criteria.

Reference: CMS MEAT Treatment Examples

Common Error: Confusing medication continuation with monitoring rather than recognizing it as active treatment.

Question 216 (Multiple Choice) Documentation stating "history of MI, no current cardiac symptoms" would:

A) Support HCC payment for old MI B) Not support any HCC payment C) Require additional cardiac evaluation D) Need specialist confirmation

Answer: B.

Historical mention without current monitoring, evaluation, assessment, or treatment does not meet MEAT criteria for HCC payment eligibility.

Reference: CMS Historical Condition Documentation

Common Error: Assuming historical conditions automatically qualify without ongoing clinical relevance.

Question 217 (Scenario-Based) A physician documents "DM type 2 - A1c 7.2%, will recheck in 3 months." Which MEAT elements are demonstrated?

A) Monitoring only B) Monitoring and Assessing C) Evaluating and Treating D) Monitoring and Evaluating

Answer: D.

"A1c 7.2%" represents Monitoring through lab review, and the clinical interpretation that warrants rechecking represents Evaluating the condition status.

Reference: CMS MEAT Laboratory Monitoring Examples

Common Error: Not recognizing lab result review and clinical interpretation as separate MEAT elements.

Question 218 (Multiple Choice) Which documentation would best support ongoing HCC eligibility for chronic kidney disease?

A) "History of CKD" B) "CKD stage 3 - monitoring creatinine levels" C) "Kidney function was abnormal last year" D) "Patient reports kidney problems"

Answer: B.

Current clinical monitoring of kidney function with specific staging demonstrates ongoing management meeting MEAT criteria.

Reference: CMS Chronic Disease Documentation Examples

Common Error: Accepting historical references without evidence of ongoing clinical management.

Question 219 (Multiple Choice) The primary purpose of MEAT criteria is to ensure:

A) Complete documentation of all patient conditions B) Conditions are clinically relevant to current care C) All treatments are evidence-based D) Providers bill appropriately

Answer: B.

MEAT criteria ensure that only conditions with ongoing clinical relevance and management generate risk adjustment payments.

Reference: CMS MEAT Criteria Purpose Statement

Common Error: Viewing MEAT as documentation completeness rather than clinical relevance criteria.

Question 220 (Select All That Apply) Which activities demonstrate "Evaluating" under MEAT criteria? (Select all that apply)

A) Reviewing response to treatment B) Assessing symptom changes C) Determining disease progression D) Scheduling routine appointments E) Comparing current to previous status

Answer: A, B, C, E.

Evaluating involves clinical assessment of condition status, response, and changes. Routine scheduling without clinical evaluation doesn't qualify.

Reference: CMS MEAT Evaluation Examples

Common Error: Including administrative activities rather than clinical evaluation activities.

Question 221 (Multiple Choice) Risk adjustment documentation in emergency department visits:

A) Never qualifies for HCC payments B) Can qualify if MEAT criteria are met C) Only qualifies for acute conditions D) Requires follow-up documentation

Answer: B.

ED visits can generate HCC payments if they involve acceptable providers, face-to-face encounters, and meet MEAT criteria for documented conditions.

Reference: CMS Emergency Department Documentation Guidelines

Common Error: Excluding ED encounters that meet standard documentation requirements.

Question 222 (Multiple Choice) When a physician documents "CHF exacerbation - will increase diuretic," this demonstrates:

A) Assessing only B) Treating only C) Assessing and Treating D) Monitoring and Treating

Answer: C.

Recognizing an "exacerbation" represents Assessment of condition status, and "will increase diuretic" represents Treatment modification.

Reference: CMS MEAT Multiple Element Examples

Common Error: Missing the assessment component in clinical recognition of condition changes.

Question 223 (Scenario-Based) A patient with diabetes sees an endocrinologist who documents "diabetes well-controlled, continue current regimen." Does this support HCC eligibility?

A) Yes - demonstrates ongoing treatment B) No - no changes were made C) Yes - shows monitoring and treatment D) No - lacks specific clinical details

Answer: A.

"Continue current regimen" represents ongoing Treatment, which satisfies MEAT criteria. Clinical stability doesn't negate HCC eligibility.

Reference: CMS Stable Condition Documentation

Common Error: Requiring treatment changes or clinical instability for HCC eligibility.

Question 224 (Multiple Choice) The documentation "patient reports taking medications for diabetes" would:

A) Support HCC payment if documented by acceptable provider B) Not support HCC payment - lacks clinical management C) Support HCC payment regardless of provider type D) Need additional specialist confirmation

Answer: B.

Patient self-reporting without provider clinical management, monitoring, evaluation, or assessment doesn't meet MEAT criteria.

Reference: CMS Patient Self-Report vs. Clinical Management

Common Error: Confusing patient reports with provider clinical documentation of management.

Question 225 (Multiple Choice) Which represents the strongest documentation for risk adjustment compliance?

A) Diagnosis lists without clinical context B) MEAT criteria met with supporting clinical rationale C) Maximum number of diagnoses per encounter D) Complex medical terminology usage

Answer: B.

Strong compliance documentation includes MEAT criteria satisfaction with clear clinical rationale supporting the ongoing relevance of documented conditions.

Reference: CMS Documentation Best Practices

Common Error: Focusing on quantity or complexity rather than quality and clinical relevance.

Question 226 (Select All That Apply) Acceptable face-to-face encounters include: (Select all that apply)

A) Office visits B) Hospital rounds C) Emergency department visits D) Telephone consultations E) Telehealth with video

Answer: A, B, C, E.

Face-to-face encounters include office visits, hospital rounds, ED visits, and qualifying telehealth. Telephone-only consultations don't meet face-to-face requirements.

Reference: CMS Face-to-Face Encounter Definitions

Common Error: Including telephone consultations or excluding qualifying telehealth encounters.

Question 227 (Multiple Choice) Documentation stating "diabetes - no acute issues today" would:

A) Support HCC payment through Assessment B) Not support HCC payment - lacks MEAT elements C) Support HCC payment through Monitoring D) Require additional testing to qualify

Answer: A.

"No acute issues today" represents clinical Assessment of the diabetes status, satisfying MEAT criteria.

Reference: CMS Clinical Assessment Documentation

Common Error: Not recognizing clinical status assessment as meeting MEAT criteria.

Question 228 (Multiple Choice) The minimum frequency for documenting chronic conditions in risk adjustment is:

A) Every encounter where the patient is seen B) At least once per calendar year C) Every six months D) Only when symptoms change

Answer: B.

Chronic conditions must be documented at least once per calendar year in qualifying encounters to maintain HCC eligibility.

Reference: CMS Annual Documentation Requirements

Common Error: Setting more frequent documentation requirements than regulations require.

Intermediate Level

Question 229 (Multiple Choice) Which documentation practice poses the highest compliance risk in risk adjustment?

A) Using abbreviations for common conditions B) Documenting conditions without MEAT criteria support C) Coding multiple conditions per encounter D) Using template-based documentation

Answer: B.

Documenting conditions without MEAT criteria support creates the highest compliance risk as it fails to meet basic regulatory requirements for HCC eligibility.

Reference: CMS Compliance Risk Assessment Guidelines

Common Error: Focusing on documentation style rather than regulatory compliance requirements.

Question 230 (Scenario-Based) A RADV audit finds that 25% of a plan's diabetes HCCs lack adequate supporting documentation. The plan has 2,000 diabetes members with average HCC value of $3,000. What is the potential financial impact?

A) $1,500,000 B) $1,200,000 C) $600,000 D) $750,000

Answer: A.

Financial impact: 2,000 members × 25% error rate × $3,000 = $1,500,000 potential payment adjustment.

Reference: RADV Audit Financial Impact Calculations

Common Error: Miscalculating the population affected by the error rate.

Question 231 (Multiple Choice) The most effective strategy for improving MEAT criteria documentation is:

A) Requiring longer clinical notes B) Training providers on specific MEAT examples relevant to their practice C) Using only template-based documentation D) Mandating specialist confirmation for all diagnoses

Answer: B.

Targeted training with practice-specific MEAT examples helps providers understand how to naturally incorporate compliant documentation into their workflow.

Reference: Provider Education Best Practices for Risk Adjustment

Common Error: Focusing on documentation length or templates rather than targeted education.

Question 232 (Select All That Apply) Which scenarios would likely fail RADV audit validation? (Select all that apply)

A) "Diabetes - patient reports taking metformin" without provider assessment B) "HTN - BP 140/90, will increase lisinopril dose" C) "History of MI" without current cardiac management D) "COPD - stable, continue current bronchodilator therapy" E) "Patient mentions having diabetes" without clinical evaluation

Answer: A, C, E.

Scenarios lacking provider clinical management (A, E) or historical mentions without current care (C) would likely fail validation.

Reference: RADV Audit Validation Criteria

Common Error: Not distinguishing between patient self-reporting and provider clinical management.

Question 233 (Multiple Choice) Advanced documentation analytics should focus on:

A) Length of clinical notes B) Number of diagnoses per encounter C) Presence and quality of MEAT criteria documentation D) Use of medical terminology complexity

Answer: C.

Advanced analytics should assess both the presence and quality of MEAT criteria documentation rather than superficial metrics.

Reference: Documentation Quality Analytics Framework

Common Error: Using quantitative metrics rather than qualitative compliance measures.

Question 234 (Calculation Exercise) A practice implements documentation improvement training. Pre-training MEAT criteria compliance was 70% across 1,000 chronic condition encounters. Post-training compliance increased to 85%. If each improved encounter generates $200 additional HCC value, what is the annual impact?

A) $30,000 B) $150,000 C) $200,000 D) $170,000

Answer: A.

Improvement: 85% - 70% = 15%. Additional compliant encounters: 1,000 × 15% = 150. Value: 150 × $200 = $30,000.

Reference: Documentation Improvement ROI Analysis

Common Error: Applying the percentage improvement to all encounters rather than calculating the incremental improvement.

Question 235 (Multiple Choice) The False Claims Act implications for risk adjustment documentation primarily concern:

A) Incorrect medical coding B) Submitting HCC claims without adequate supporting documentation C) Using generic documentation templates D) Billing for services not rendered

Answer: B.

FCA risk in risk adjustment primarily involves submitting claims for HCC payments when documentation doesn't adequately support the claimed conditions.

Reference: False Claims Act Risk Assessment for Risk Adjustment

Common Error: Focusing on service billing rather than the adequacy of documentation supporting HCC claims.

Question 236 (Scenario-Based) A physician queries system shows providers are asked to clarify diabetes documentation in 40% of diabetic patient encounters. This high query rate most likely indicates:

A) Appropriate thorough documentation review B) Poor baseline documentation quality requiring systematic improvement C) Overly aggressive risk adjustment practices D) Normal variation in clinical documentation

Answer: B.

A 40% query rate suggests systematic documentation quality issues requiring comprehensive provider education and workflow improvement.

Reference: Query Rate Analysis Best Practices

Common Error: Viewing high query rates as thorough review rather than indication of baseline quality issues.

Question 237 (Multiple Choice) Which documentation improvement technology is most effective for real-time compliance?

A) Automated coding software B) Clinical decision support with MEAT prompting C) Voice recognition software D) Electronic health record templates

Answer: B.

Clinical decision support that prompts providers about MEAT requirements in real-time during documentation is most effective for immediate compliance improvement.

Reference: Health Information Technology for Documentation Improvement

Common Error: Focusing on documentation capture technology rather than compliance guidance technology.

Question 238 (Select All That Apply) Effective compliance monitoring should include: (Select all that apply)

A) Random medical record audits for MEAT criteria B) Provider-specific documentation pattern analysis C) Patient satisfaction surveys D) RADV audit preparation reviews E) Diagnosis-specific documentation quality metrics

Answer: A, B, D, E.

Compliance monitoring includes medical record audits, provider pattern analysis, audit preparation, and diagnosis-specific quality metrics. Patient satisfaction isn't directly related to documentation compliance.

Reference: Comprehensive Compliance Monitoring Framework

Common Error: Including patient experience measures that don't relate to documentation compliance.

Question 239 (Multiple Choice) The most sophisticated approach to ensuring documentation supports clinical care while meeting risk adjustment requirements involves:

A) Separating clinical and risk adjustment documentation B) Integrating risk adjustment considerations into natural clinical workflow C) Using dedicated risk adjustment encounters D) Requiring risk adjustment specialists to review all documentation

Answer: B.

Integration into natural clinical workflow ensures documentation authentically reflects clinical care while meeting compliance requirements.

Reference: Integrated Clinical Documentation Best Practices

Common Error: Creating artificial separation between clinical care and risk adjustment documentation.

Question 240 (Multiple Choice) Natural Language Processing (NLP) applications for documentation compliance are most effective for:

A) Automatically generating compliant documentation B) Identifying potential documentation gaps and MEAT deficiencies C) Replacing physician documentation review D) Creating standardized clinical notes

Answer: B.

NLP excels at identifying patterns indicating potential documentation gaps or MEAT deficiencies that warrant human review and improvement.

Reference: NLP Applications in Documentation Quality

Common Error: Expecting NLP to replace human judgment rather than identify opportunities for improvement.

Question 241 (Scenario-Based) A health plan discovers that cardiologists document heart failure management in 95% of encounters, while primary care physicians document it in only 60% of similar patient encounters. What is the most appropriate intervention?

A) Require all heart failure patients to see cardiologists B) Implement heart failure-specific documentation training for PCPs C) Automatically add heart failure documentation to PCP notes D) Transfer all heart failure management to cardiology

Answer: B.

Targeted education helps PCPs recognize and document heart failure management activities they may already be performing but not adequately capturing.

Reference: Provider-Specific Documentation Improvement

Common Error: Changing care delivery rather than improving documentation of existing appropriate care.

Question 242 (Multiple Choice) Risk adjustment documentation audits should primarily assess:

A) Grammar and spelling accuracy B) Clinical appropriateness and MEAT criteria compliance C) Note length and detail level D) Use of approved medical terminology

Answer: B.

Audits should focus on whether documented conditions are clinically appropriate and meet MEAT criteria for ongoing relevance to patient care.

Reference: Risk Adjustment Documentation Audit Standards

Common Error: Focusing on documentation format rather than clinical appropriateness and compliance.

Question 243 (Calculation Exercise) A health plan's documentation audit reviews 500 HCC encounters and finds:

- 400 fully support HCC claims
- 75 have minor MEAT deficiencies
- 25 have major documentation deficiencies

What is the audit validation rate?

A) 80% B) 85% C) 95% D) 90%

Answer: A.

Fully supported encounters: 400. Validation rate: 400/500 = 80%. Minor deficiencies may still support claims depending on specific criteria.

Reference: Documentation Audit Validation Metrics

Common Error: Including encounters with minor deficiencies as fully validated without considering specific validation standards.

Question 244 (Multiple Choice) The Office of Inspector General (OIG) focuses on risk adjustment documentation that:

A) Uses too much medical terminology B) Lacks adequate clinical support for claimed conditions C) Is too brief or concise D) Doesn't follow specific formatting standards

Answer: B.

OIG enforcement focuses on whether documentation adequately supports the clinical reality and ongoing management of conditions claimed for risk adjustment payments.

Reference: OIG Risk Adjustment Enforcement Priorities

Common Error: Focusing on documentation style rather than adequacy of clinical support for claimed conditions.

Question 245 (Select All That Apply) Red flags for risk adjustment documentation compliance include: (Select all that apply)

A) Sudden increases in HCC capture rates without corresponding clinical programs B) Documentation changes that exactly mirror risk adjustment requirements without clinical rationale C) High use of medical abbreviations D) Provider documentation patterns that significantly deviate from clinical norms E) Regular use of documentation templates

Answer: A, B, D.

Red flags include sudden unexplained increases, documentation changes lacking clinical rationale, and patterns deviating from clinical norms. Abbreviations and templates aren't inherently problematic.

Reference: Risk Adjustment Compliance Red Flag Analysis

Common Error: Including documentation style elements rather than patterns suggesting inappropriate practice.

Question 246 (Multiple Choice) Effective provider education on risk adjustment documentation should emphasize:

A) Maximizing the number of documented conditions B) Documenting conditions that genuinely affect patient care and management C) Using complex medical terminology D) Following rigid documentation templates

Answer: B.

Education should focus on authentic documentation of conditions that genuinely affect patient care, ensuring natural integration with clinical practice.

Reference: Provider Education Best Practices

Common Error: Emphasizing documentation quantity or complexity rather than clinical authenticity.

Question 247 (Scenario-Based) A practice's documentation review finds providers frequently document "diabetes - stable" without additional clinical context. What improvement strategy would be most effective?

A) Require longer diabetes documentation B) Provide specific examples of MEAT-compliant diabetes documentation C) Mandate specialist diabetes consultations D) Use automated diabetes documentation templates

Answer: B.

Specific examples help providers understand how to naturally incorporate MEAT elements into their existing documentation style.

Reference: Targeted Provider Education Strategies

Common Error: Requiring length or specialist involvement rather than improving existing documentation quality.

Question 248 (Multiple Choice) The most defensible documentation for risk adjustment purposes includes:

A) Extensive copying from previous notes B) Original clinical thinking and decision-making for each encounter C) Standardized template responses for all conditions D) Maximum technical medical terminology

Answer: B.

Original clinical thinking demonstrating current evaluation and management provides the strongest defense for HCC claims during audits.

Reference: Defensible Documentation Standards

Common Error: Relying on copied content or templates rather than original clinical assessment.

Question 249 (Multiple Choice) When implementing documentation improvement programs, the most important success metric is:

A) Increased number of diagnoses per encounter B) Improved audit validation rates for HCC claims C) Faster note completion times D) Higher patient satisfaction scores

Answer: B.

Improved audit validation rates demonstrate that documentation improvements actually enhance compliance and defensibility of HCC claims.

Reference: Documentation Improvement Program Metrics

Common Error: Focusing on quantity or speed rather than quality and compliance outcomes.

Question 250 (Select All That Apply) Machine learning applications in documentation compliance can effectively: (Select all that apply)

A) Identify patterns suggesting potential MEAT deficiencies B) Flag unusual documentation patterns for review C) Automatically generate compliant clinical documentation D) Predict which encounters may fail RADV audits E) Replace clinical judgment in documentation decisions

Answer: A, B, D.

ML can identify patterns, flag anomalies, and predict audit risk, but cannot generate authentic clinical documentation or replace clinical judgment.

Reference: Machine Learning in Documentation Quality Assurance

Common Error: Expecting ML to replace human clinical documentation and judgment rather than augment quality assurance.

Question 251 (Multiple Choice) The primary difference between clinical documentation improvement (CDI) and risk adjustment documentation is:

A) CDI focuses on inpatient settings only B) Risk adjustment requires annual documentation while CDI is encounter-specific C) CDI emphasizes coding accuracy while risk adjustment emphasizes ongoing clinical relevance D) They are essentially the same process

Answer: C.

CDI typically focuses on accurate coding and capture, while risk adjustment documentation emphasizes ongoing clinical relevance and management (MEAT criteria).

Reference: CDI vs. Risk Adjustment Documentation Comparison

Common Error: Treating CDI and risk adjustment documentation as identical processes with different focus areas.

Question 252 (Scenario-Based) A physician documents "COPD - patient feels short of breath, increased rescue inhaler use, will start oral steroids." How many MEAT elements are demonstrated?

A) One - Treating B) Two - Evaluating and Treating
C) Three - Monitoring, Evaluating, and Treating D) Four - All MEAT elements

Answer: C.

"Feels short of breath" = Monitoring symptoms, "increased rescue inhaler use" = Evaluating condition status, "will start oral steroids" = Treating.

Reference: MEAT Elements Analysis Examples

Common Error: Missing the monitoring component in symptom documentation or not recognizing evaluation of medication usage patterns.

Question 253 (Multiple Choice) Risk adjustment documentation in skilled nursing facilities (SNF):

A) Cannot generate HCC payments B) Must meet the same MEAT criteria as other settings C) Only applies to Medicare Part A stays D) Requires different documentation standards

Answer: B.

SNF documentation must meet the same MEAT criteria and acceptable provider requirements as other healthcare settings for risk adjustment eligibility.

Reference: SNF Risk Adjustment Documentation Requirements

Common Error: Assuming different settings have different MEAT criteria or assuming SNF encounters don't qualify.

Question 254 (Multiple Choice) The most effective approach to reducing documentation compliance risk is:

A) Hiring more compliance staff B) Implementing comprehensive provider education and real-time decision support C) Reducing the number of documented conditions D) Requiring legal review of all documentation

Answer: B.

Comprehensive education combined with real-time decision support prevents compliance issues at the source rather than detecting them after occurrence.

Reference: Proactive Compliance Risk Reduction Strategies

Common Error: Focusing on detection and review rather than prevention through education and support.

Question 255 (Calculation Exercise) A health plan's compliance program costs $500,000 annually and prevents an estimated $2.5 million in potential RADV audit adjustments. The program also generates $1.2 million in improved HCC capture. What is the ROI?

A) 500% B) 640% C) 240% D) 740%

Answer: B.

Total benefit: $2.5M (prevented adjustments) + $1.2M (improved capture) = $3.7M. ROI: ($3.7M - $0.5M) / $0.5M = 640%.

Reference: Compliance Program ROI Analysis

Common Error: Not including all benefit categories or miscalculating return ratios.

Question 256 (Multiple Choice) Documentation stating "continue monitoring diabetes labs" represents which MEAT element?

A) Monitoring B) Evaluating
C) Assessing D) Treating

Answer: D.

"Continue monitoring" represents ongoing Treatment through laboratory surveillance, not the monitoring activity itself.

Reference: MEAT Treatment vs. Monitoring Distinction

Common Error: Confusing the treatment plan for monitoring with the monitoring activity itself.

Question 257 (Select All That Apply) Best practices for risk adjustment documentation queries include: (Select all that apply)

A) Leading questions that suggest specific diagnoses B) Open-ended questions about clinical findings C) Multiple choice options for diagnosis selection D) Requests for clarification of documented clinical evidence E) Questions that facilitate clinical decision-making

Answer: B, D, E.

Effective queries use open-ended questions, seek clarification of existing evidence, and facilitate clinical thinking rather than suggesting specific diagnoses.

Reference: Clinical Documentation Query Best Practices

Common Error: Using leading questions or multiple choice formats that suggest specific clinical conclusions.

Question 258 (Multiple Choice) The strongest documentation to support chronic kidney disease HCC claims includes:

A) "Patient has kidney problems" B) "CKD stage 3 - monitoring creatinine, eGFR stable at 45" C) "History of elevated creatinine" D) "Kidney function abnormal per labs"

Answer: B.

Specific staging with current monitoring and clinical values demonstrates ongoing management meeting MEAT criteria.

Reference: Chronic Disease Documentation Best Practices

Common Error: Accepting general references without specific clinical management documentation.

Question 259 (Scenario-Based) A health plan's documentation audit reveals that 80% of mental health HCCs are documented by primary care physicians with limited psychiatric detail. Is this a compliance concern?

A) Yes - mental health conditions require specialist documentation B) No - if MEAT criteria are met by qualified providers C) Yes - primary care physicians cannot document psychiatric conditions D) No - mental health HCCs don't require specific documentation standards

Answer: B.

Any qualified provider can document mental health conditions for risk adjustment if MEAT criteria are met, regardless of specialty.

Reference: Provider Type Requirements for Mental Health Documentation

Common Error: Assuming specialty-specific documentation requirements that don't exist in risk adjustment regulations.

Question 260 (Multiple Choice) Advanced documentation analytics should identify:

A) Providers who document too many conditions B) Documentation patterns inconsistent with clinical evidence and best practices C) Notes that are too long or detailed D) Use of medical abbreviations

Answer: B.

Advanced analytics should identify patterns that suggest documentation may not align with clinical evidence or appropriate practice standards.

Reference: Advanced Documentation Pattern Analysis

Common Error: Focusing on quantity or style rather than clinical appropriateness and evidence alignment.

Question 261 (Multiple Choice) The documentation "diabetes well-controlled, A1c 6.8%, continue current therapy" demonstrates:

A) Monitoring only B) Evaluating only C) Monitoring, Evaluating, and Treating D) Assessing and Treating

Answer: C.

"A1c 6.8%" = Monitoring, "well-controlled" = Evaluating status, "continue current therapy" = Treating.

Reference: Complex MEAT Documentation Analysis

Common Error: Missing the evaluation component in clinical status assessment.

Question 262 (Select All That Apply) Documentation compliance programs should include: (Select all that apply)

A) Regular provider education updates B) Pre-documentation query processes C) Post-encounter documentation audits D) Real-time clinical decision support E) Automated documentation generation

Answer: A, C, D.

Effective programs include education, audits, and real-time support. Pre-documentation queries aren't standard practice, and automated generation compromises clinical authenticity.

Reference: Comprehensive Documentation Compliance Program Elements

Common Error: Including inappropriate pre-documentation or automated processes that compromise clinical integrity.

Question 263 (Multiple Choice) When providers document "patient reports taking diabetes medication," this:

A) Meets MEAT criteria if documented by qualified provider B) Does not meet MEAT criteria without provider clinical management C) Requires specialist confirmation to qualify D) Automatically qualifies for HCC payment

Answer: B.

Patient self-reporting without provider clinical evaluation, monitoring, assessment, or treatment decisions doesn't meet MEAT criteria.

Reference: Patient Self-Report vs. Provider Management Documentation

Common Error: Confusing patient reports with provider clinical management activities.

Question 264 (Scenario-Based) A provider documents "hypertension - BP elevated at 150/95, patient admits non-compliance with medications, counseled on importance of adherence." Which MEAT elements are present?

A) Monitoring and Treating B) Monitoring, Evaluating, and Treating C) Assessing and Treating D) All four MEAT elements

Answer: B.

"BP elevated at 150/95" = Monitoring, "admits non-compliance" = Evaluating response to treatment, "counseled" = Treating through education.

Reference: Comprehensive MEAT Documentation Examples

Common Error: Not recognizing patient education and counseling as treatment interventions.

Question 265 (Multiple Choice) The most challenging aspect of MEAT criteria compliance is:

A) Understanding what qualifies as face-to-face encounters B) Distinguishing between historical conditions and ongoing clinical management C) Identifying acceptable provider types D) Documenting acute vs. chronic conditions

Answer: B.

The distinction between historical mentions and ongoing clinical relevance requiring current management is often the most challenging compliance aspect.

Reference: Common MEAT Criteria Compliance Challenges

Common Error: Treating all historical condition mentions as automatically qualifying for ongoing HCC payments.

Question 266 (Multiple Choice) Documentation improvement programs show the greatest ROI when they focus on:

A) Conditions with the highest HCC values B) Conditions most commonly documented by providers C) Conditions with the largest gaps between clinical reality and documentation D) Acute conditions requiring immediate treatment

Answer: C.

Programs targeting gaps between clinical reality and documentation address authentic care that's not being captured, providing both clinical and financial benefits.

Reference: Documentation Improvement Program ROI Optimization

Common Error: Focusing only on high-value conditions rather than areas with the largest improvement opportunities.

Question 267 (Calculation Exercise) A practice reviews 200 diabetes encounters and finds MEAT criteria compliance rates of:

- Monitoring: 85%
- Evaluating: 70%
- Assessing: 60%
- Treating: 90%

What percentage of encounters document at least one MEAT element?

A) 76.25% (average) B) 95% (estimated based on overlap) C) 90% (highest single element) D) Cannot determine without encounter-level data

Answer: D.

Cannot determine the percentage with at least one MEAT element without knowing how elements overlap in individual encounters.

Reference: MEAT Criteria Statistical Analysis

Common Error: Assuming averages or maximums represent the overlap scenario.

Question 268 (Multiple Choice) The future of risk adjustment documentation will likely emphasize:

A) Longer, more detailed clinical notes B) Integration with real-time clinical decision support and AI assistance C) Standardized documentation templates D) Separation of clinical and risk adjustment documentation

Answer: B.

Future development will likely integrate intelligent decision support that helps providers document authentic clinical care that naturally meets compliance requirements.

Reference: Future Trends in Clinical Documentation Technology

Common Error: Expecting increased length or standardization rather than intelligent assistance that maintains clinical authenticity.

Advanced Level

Question 269 (Multiple Choice) The most sophisticated risk adjustment documentation compliance programs utilize:

A) Random audit sampling across all providers B) Predictive analytics to identify high-risk documentation patterns before audit C) Mandatory documentation templates for all encounters D) Retrospective correction of documentation deficiencies

Answer: B.

Advanced programs use predictive analytics to identify potential compliance risks proactively rather than waiting for audit findings.

Reference: Predictive Compliance Analytics in Risk Adjustment

Common Error: Relying on reactive rather than predictive compliance approaches.

Question 270 (Scenario-Based) A health plan's machine learning model identifies that Provider A documents diabetes complications in 85% of diabetic patients while Provider B documents them in 35% of similar patients. Both providers' documentation passes manual review for clinical appropriateness. What is the most sophisticated interpretation?

A) Provider A is likely overcoding B) Provider B is likely undercoding C) The variation represents different patient populations, clinical practices, or documentation styles, all potentially appropriate D) Both providers need standardization training

Answer: C.

Sophisticated analysis recognizes that variation can represent legitimate differences in patient populations, clinical assessment capabilities, or documentation practices when both pass clinical appropriateness review.

Reference: Advanced Provider Variation Analysis

Common Error: Assuming variation automatically indicates inappropriate practice rather than legitimate clinical or practice differences.

Question 271 (Multiple Choice) Advanced natural language processing (NLP) for MEAT criteria assessment should prioritize:

A) Keyword identification for MEAT elements B) Clinical context analysis to determine genuine management activities C) Maximum extraction of all possible medical terms D) Standardization of clinical language

Answer: B.

Advanced NLP analyzes clinical context to distinguish genuine management activities meeting MEAT criteria from superficial mentions or historical references.

Reference: Advanced NLP in Clinical Documentation Assessment

Common Error: Focusing on keyword detection rather than contextual clinical relevance analysis.

Question 272 (Calculation Exercise) A health plan implements advanced documentation analytics that identifies potential MEAT deficiencies in real-time. The system flags 3,000 encounters annually for review. Manual validation confirms 2,100 truly have deficiencies. Intervention prevents 85% of flagged deficiencies. If each prevented deficiency saves $1,500 in potential audit adjustments, what is the annual value?

A) $2,677,500 B) $3,825,000 C) $4,500,000 D) $2,295,000

Answer: A.

Confirmed deficiencies: 2,100. Prevented deficiencies: 2,100 × 85% = 1,785. Value: 1,785 × $1,500 = $2,677,500.

Reference: Advanced Analytics ROI Calculation

Common Error: Not accounting for the manual validation step or the intervention effectiveness rate.

Question 273 (Select All That Apply) Sophisticated documentation compliance monitoring should include: (Select all that apply)

A) Real-time MEAT criteria validation during documentation B) Provider-specific risk profiling based on documentation patterns C) Predictive modeling for RADV audit risk assessment D) Automated documentation correction systems E) Clinical appropriateness verification using AI

Answer: A, B, C, E.

Advanced monitoring includes real-time validation, risk profiling, predictive modeling, and AI-assisted clinical appropriateness verification. Automated correction compromises clinical authenticity.

Reference: Next-Generation Documentation Compliance Systems

Common Error: Including automated correction that compromises the authenticity of clinical documentation.

Question 274 (Multiple Choice) The most advanced approach to managing documentation for conditions with complex MEAT requirements involves:

A) Standardized templates for all complex conditions B) Clinical decision support that adapts to provider workflow and specialty C) Requiring specialist consultation for all complex conditions D) Automated MEAT element insertion

Answer: B.

Adaptive clinical decision support integrates naturally with provider workflow and specialty-specific practices while maintaining clinical authenticity.

Reference: Adaptive Clinical Decision Support Systems

Common Error: Using rigid templates or automated insertion rather than adaptive, workflow-integrated support.

Question 275 (Scenario-Based) A health plan's advanced analytics identify that mental health HCCs documented by primary care physicians have a 15% higher RADV audit failure rate than those documented by psychiatrists, despite similar clinical presentation complexity. What is the most sophisticated intervention?

A) Require psychiatric consultation for all mental health diagnoses B) Implement PCP-specific mental health documentation training with psychiatrist consultation support C) Transfer all mental health management to psychiatrists D) Accept the higher failure rate as unavoidable

Answer: B.

Sophisticated intervention maintains appropriate care delivery while providing targeted support to address documentation quality differences.

Reference: Specialty-Specific Documentation Enhancement Strategies

Common Error: Changing appropriate care delivery models rather than enhancing documentation capabilities.

Question 276 (Multiple Choice) Machine learning applications for predicting RADV audit outcomes should focus on:

A) Simple pattern matching of documentation keywords B) Complex analysis of clinical context, documentation quality, and provider patterns C) Random sampling of documentation elements D) Historical audit results only

Answer: B.

Advanced ML analyzes multiple dimensions including clinical context, documentation quality patterns, and provider-specific factors to predict audit risk.

Reference: Machine Learning for Audit Risk Prediction

Common Error: Using simple pattern matching rather than comprehensive multifactor analysis.

Question 277 (Multiple Choice) The most sophisticated documentation quality metrics for risk adjustment include:

A) Average number of conditions documented per encounter B) Clinical appropriateness scores adjusted for documentation complexity and MEAT criteria compliance C) Speed of documentation completion D) Percentage of encounters with at least five diagnoses

Answer: B.

Advanced metrics assess clinical appropriateness while accounting for documentation complexity and regulatory compliance rather than simple quantity measures.

Reference: Advanced Documentation Quality Measurement

Common Error: Using quantitative metrics rather than quality and appropriateness measures.

Question 278 (Scenario-Based) A health plan's predictive model identifies members likely to develop diabetes complications within 12 months with 90% accuracy. How should this inform documentation strategy?

A) Pre-code complications for high-risk members B) Implement enhanced screening and monitoring protocols for high-risk members C) Assume complications will develop and adjust risk scores D) Focus documentation efforts only on currently complicated diabetes

Answer: B.

Predictive insights should inform clinical care improvements that may lead to earlier appropriate identification and documentation of developing conditions.

Reference: Predictive Analytics in Clinical Care Enhancement

Common Error: Using predictions to directly influence coding rather than clinical care that leads to appropriate documentation.

Question 279 (Multiple Choice) Advanced risk adjustment documentation programs integrate with:

A) Coding systems only B) Clinical care delivery, quality measurement, and population health management C) Billing systems exclusively D) Administrative reporting functions

Answer: B.

Sophisticated programs integrate documentation improvement with broader clinical care, quality initiatives, and population health strategies.

Reference: Integrated Healthcare Management Systems

Common Error: Limiting integration to administrative or coding functions rather than comprehensive clinical integration.

Question 280 (Calculation Exercise) A health plan's advanced documentation system uses AI to identify potential MEAT deficiencies. The system reviews 50,000 encounters annually, flags 8,000 for potential issues, and manual review confirms 6,400 true deficiencies. Intervention corrects 90% of confirmed deficiencies. If each correction prevents $800 in potential audit adjustment, what is the system's annual value?

A) $6,400,000 B) $5,760,000 C) $4,608,000 D) $3,840,000

Answer: C.

Confirmed deficiencies: 6,400. Corrected deficiencies: 6,400 × 90% = 5,760. Value: 5,760 × $800 = $4,608,000.

Reference: AI-Powered Documentation System ROI

Common Error: Not accounting for manual confirmation or intervention effectiveness rates.

Question 281 (Select All That Apply) Next-generation documentation compliance should address: (Select all that apply)

A) Real-time clinical decision support during documentation B) Predictive identification of high-audit-risk encounters C) Integration with social determinants of health data D) Automated generation of compliant documentation E) Provider-specific adaptive learning systems

Answer: A, B, E.

Next-generation systems provide real-time support, predictive risk identification, and adaptive learning. SDOH integration and automated generation don't directly enhance documentation compliance.

Reference: Future Documentation Compliance Technology

Common Error: Including automated generation that compromises clinical authenticity or SDOH data that doesn't affect documentation requirements.

Question 282 (Multiple Choice) The most advanced approach to training providers on MEAT criteria involves:

A) Annual classroom-based training sessions B) Real-time, case-specific coaching integrated into workflow with immediate feedback C) Online modules with standardized examples D) Written policy distribution

Answer: B.

Advanced training provides real-time, contextual coaching during actual clinical work with immediate feedback on documentation decisions.

Reference: Advanced Provider Education Technology

Common Error: Relying on traditional classroom or online training rather than integrated workflow education.

Question 283 (Multiple Choice) Sophisticated analysis of documentation patterns should identify:

A) Providers who document the most conditions B) Documentation that lacks clinical coherence or appears disconnected from patient care C) Longest clinical notes D) Most complex medical terminology usage

Answer: B.

Advanced analysis identifies documentation that may lack clinical coherence or authentic connection to patient care, indicating potential compliance risks.

Reference: Advanced Documentation Pattern Analysis

Common Error: Focusing on quantity or complexity rather than clinical coherence and authenticity.

Question 284 (Scenario-Based) A health plan's AI system identifies that Provider X's documentation pattern changed significantly after risk adjustment training, with diabetes complication documentation increasing from 20% to 75% of diabetic patients. Clinical review confirms appropriateness. What is the most sophisticated interpretation?

A) The provider is likely gaming the system B) The training successfully helped the provider recognize and document previously unidentified clinical conditions C) The change is too dramatic to be legitimate D) An audit is immediately warranted

Answer: B.

Sophisticated analysis recognizes that effective training can reveal significant gaps in prior documentation of legitimate clinical conditions.

Reference: Training Impact Assessment in Documentation Improvement

Common Error: Assuming dramatic improvements automatically indicate inappropriate activity rather than successful education addressing prior gaps.

Question 285 (Multiple Choice) Advanced clinical decision support for MEAT criteria should:

A) Automatically insert MEAT elements into documentation B) Provide contextual suggestions based on documented clinical findings C) Require providers to select from predetermined MEAT options D) Replace clinical judgment in documentation decisions

Answer: B.

Advanced systems provide contextual suggestions that help providers recognize how their clinical findings and decisions can be documented to meet MEAT criteria.

Reference: Contextual Clinical Decision Support Design

Common Error: Using automated insertion or predetermined options rather than intelligent contextual guidance.

Question 286 (Multiple Choice) The future evolution of documentation compliance will likely emphasize:

A) Increased documentation requirements and complexity B) Intelligent assistance that maintains clinical authenticity while ensuring compliance C) Standardization of all clinical documentation D) Elimination of provider documentation requirements

Answer: B.

Future systems will likely provide intelligent assistance that helps providers document authentic clinical care in ways that naturally meet compliance requirements.

Reference: Future Trends in Clinical Documentation Technology

Common Error: Expecting increased complexity or standardization rather than intelligent assistance that maintains authenticity.

Question 287 (Select All That Apply) Advanced documentation analytics for risk adjustment should monitor: (Select all that apply)

A) Real-time compliance with MEAT criteria during documentation B) Provider-specific learning curves after training interventions C) Correlation between documentation changes and clinical outcomes D) Patient satisfaction with documentation discussions E) Predictive models for documentation sustainability

Answer: A, B, C, E.

Advanced analytics monitor real-time compliance, learning effectiveness, clinical correlations, and sustainability predictions. Patient satisfaction with documentation discussions isn't typically measurable or relevant.

Reference: Comprehensive Documentation Analytics Framework

Common Error: Including patient satisfaction metrics that don't relate to documentation compliance or quality.

Question 288 (Multiple Choice) The most sophisticated approach to managing documentation for rare or complex conditions involves:

A) Requiring specialist consultation for all rare conditions B) Using AI-powered clinical decision support with specialty-specific guidance C) Accepting lower documentation quality for complex conditions D) Standardizing documentation regardless of condition complexity

Answer: B.

AI-powered support can provide specialty-specific guidance for complex conditions while maintaining documentation quality and clinical appropriateness.

Reference: AI-Assisted Specialty Documentation Support

Common Error: Requiring specialist involvement or accepting lower standards rather than providing intelligent specialized support.

Question 289 (Scenario-Based) A health plan's documentation compliance program achieves 95% RADV audit validation rates but costs $2 million annually. A competing approach costs $3.5 million but achieves 98% validation. The plan has $100 million in annual HCC payments. Which approach provides better value?

A) Current approach - lower cost with acceptable results B) Alternative approach - higher validation worth the additional cost C) Need more information about potential audit adjustment amounts D) Both approaches are equally valuable

Answer: C.

Cannot determine value without knowing the potential financial impact of the 3% validation difference on $100M in HCC payments.

Reference: Complex ROI Analysis for Compliance Programs

Common Error: Making decisions based on percentages without considering the financial magnitude of the differences.

Question 290 (Multiple Choice) Advanced documentation improvement programs demonstrate success through:

A) Increased number of diagnoses per encounter B) Sustained improvement in audit validation rates and clinical outcome correlation C) Faster documentation completion times D) Higher provider satisfaction scores

Answer: B.

True success includes both compliance improvement and correlation with clinical outcomes, demonstrating authentic care documentation rather than just compliance gaming.

Reference: Comprehensive Program Success Metrics

Common Error: Focusing on quantity or efficiency metrics rather than quality and authenticity measures.

Question 291 (Multiple Choice) The integration of blockchain technology in documentation compliance would most appropriately address:

A) Real-time MEAT criteria validation B) Documentation immutability and audit trail integrity C) Clinical decision support enhancement D) Provider education delivery

Answer: B.

Blockchain technology is best suited for ensuring documentation integrity and immutable audit trails rather than real-time validation or decision support.

Reference: Blockchain Applications in Healthcare Documentation

Common Error: Applying blockchain to problems better solved by other technologies rather than its core strength in immutability.

Question 292 (Calculation Exercise) A health plan implements a comprehensive documentation program with these annual costs and benefits:

- Program cost: $1.8M
- Prevented RADV adjustments: $4.2M
- Improved HCC capture: $2.8M
- Reduced compliance staff needs: $0.6M savings
- Provider efficiency gains: $1.1M

What is the net ROI?

A) 344% B) 383% C) 478% D) 267%

Answer: B.

Total benefits: $4.2M + $2.8M + $0.6M + $1.1M = $8.7M. Net benefit: $8.7M - $1.8M = $6.9M. ROI: $6.9M ÷ $1.8M = 383%.

Reference: Comprehensive Program ROI Analysis

Common Error: Not including all benefit categories or miscalculating the ROI formula.

Question 293 (Multiple Choice) Advanced documentation systems should prioritize:

A) Maximum automation to reduce provider burden B) Clinical authenticity with intelligent compliance assistance C) Standardization across all provider types and specialties D) Integration with billing systems for revenue optimization

Answer: B.

Advanced systems maintain clinical authenticity while providing intelligent assistance that helps providers meet compliance requirements naturally.

Reference: Optimal Documentation System Design Principles

Common Error: Prioritizing automation or revenue optimization over clinical authenticity and appropriate compliance support.

Question 294 (Scenario-Based) A health plan's advanced analytics identify that documentation quality varies significantly based on EHR system used, with System A showing 20% better MEAT compliance than System B across similar providers. What is the most sophisticated response?

A) Mandate all providers use System A B) Develop System B-specific documentation enhancement tools and training C) Accept the variation as unavoidable D) Provide financial incentives for System A adoption

Answer: B.

Sophisticated response addresses the specific challenges of each system rather than forcing uniformity or accepting suboptimal performance.

Reference: EHR-Specific Documentation Optimization

Common Error: Mandating system changes rather than optimizing performance within existing systems.

Question 295 (Multiple Choice) The most advanced metric for assessing documentation program effectiveness is:

A) Provider satisfaction with documentation tools B) Correlation between documentation improvements and both audit outcomes and clinical quality measures C) Speed of documentation completion D) Number of MEAT elements documented per encounter

Answer: B.

Advanced assessment measures both compliance outcomes and clinical quality correlation, ensuring documentation improvements reflect authentic care enhancement.

Reference: Advanced Program Effectiveness Measurement

Common Error: Using single-dimension metrics rather than comprehensive quality and outcome correlation measures.

Question 296 (Select All That Apply) Future documentation compliance systems should integrate: (Select all that apply)

A) Predictive analytics for audit risk assessment B) Real-time clinical decision support C) Automated documentation generation D) Provider-specific adaptive learning algorithms E) Population health trend analysis

Answer: A, B, D.

Future systems should integrate predictive analytics, real-time support, and adaptive learning. Automated generation compromises authenticity, and population trends don't directly affect individual documentation.

Reference: Future Documentation System Integration Requirements

Common Error: Including automated generation that compromises clinical authenticity or population-level data that doesn't affect individual documentation decisions.

Question 297 (Multiple Choice) The most sophisticated approach to documentation compliance education involves:

A) Annual mandatory training for all providers B) Continuous, personalized learning based on individual provider patterns and performance C) Standardized online modules for all specialties D) Compliance-focused training separated from clinical education

Answer: B.

Advanced education provides continuous, personalized learning that adapts to individual provider needs and performance patterns.

Reference: Personalized Provider Education Systems

Common Error: Using standardized approaches rather than personalized, continuous learning systems.

Question 298 (Multiple Choice) Advanced documentation validation should emphasize:

A) Keyword presence indicating MEAT elements B) Clinical coherence and authentic care relationship assessment C) Compliance with documentation templates D) Similarity to other providers' documentation patterns

Answer: B.

Advanced validation assesses whether documentation reflects authentic clinical care relationships and coherent management rather than superficial compliance indicators.

Reference: Advanced Clinical Documentation Validation

Common Error: Focusing on superficial compliance indicators rather than authentic clinical care assessment.

Question 299 (Scenario-Based) A health plan's AI system identifies potential documentation improvements that could generate $5M annually in additional HCC capture. However, clinical review suggests only 60% represents authentic clinical conditions not previously documented. What is the ethical approach?

A) Implement all AI suggestions to maximize revenue B) Implement only the 60% that represents authentic clinical documentation improvement C) Reject all AI suggestions to avoid compliance risk D) Implement suggestions gradually to avoid audit attention

Answer: B.

Ethical approach focuses on authentic clinical documentation improvement rather than maximizing revenue through questionable documentation enhancement.

Reference: Ethical Documentation Improvement Practices

Common Error: Prioritizing revenue maximization over clinical authenticity and appropriate documentation.

Question 300 (Multiple Choice) The ultimate goal of advanced documentation compliance programs should be:

A) Maximum HCC revenue generation B) Perfect RADV audit validation rates C) Authentic clinical documentation that naturally meets compliance requirements while supporting quality care D) Standardized documentation across all providers

Answer: C.

The ultimate goal is authentic clinical documentation that naturally supports both compliance and quality care rather than artificial compliance optimization.

Reference: Documentation Program Strategic Objectives

Common Error: Focusing on revenue or compliance metrics rather than authentic clinical care documentation that naturally achieves compliance.

Medical Terminology

Foundation Level

Question 301 (Multiple Choice) What does the suffix "-itis" indicate?

A) Enlargement B) Inflammation C) Surgical removal D) Disease condition

Answer: B.

The suffix "-itis" indicates inflammation, as seen in conditions like arthritis (joint inflammation) or gastritis (stomach inflammation).

Reference: Medical Terminology Fundamentals

Common Error: Confusing "-itis" with other condition-indicating suffixes like "-osis" or "-pathy."

Question 302 (Multiple Choice) The prefix "hyper-" means:

A) Below normal B) Above normal C) Around or surrounding D) Through or across

Answer: B.

"Hyper-" means above normal or excessive, as in hypertension (high blood pressure) or hyperglycemia (high blood sugar).

Reference: Medical Terminology Prefix Guide

Common Error: Confusing "hyper-" with "hypo-" which means below normal.

Question 303 (Select All That Apply) Which terms relate to the cardiovascular system? (Select all that apply)

A) Cardiology B) Nephrology C) Myocardial D) Pulmonary E) Vascular

Answer: A, C, E.

Cardiology (heart study), myocardial (heart muscle), and vascular (blood vessels) relate to the cardiovascular system. Nephrology relates to kidneys, pulmonary to lungs.

Reference: Medical Terminology by Body System

Common Error: Including terms from other body systems that may affect cardiovascular health but aren't primarily cardiovascular terms.

Question 304 (Multiple Choice) The term "nephropathy" refers to:

A) Kidney inflammation B) Kidney disease C) Kidney enlargement D) Kidney surgical removal

Answer: B.

"Nephropathy" means kidney disease, with "nephro-" referring to kidney and "-pathy" meaning disease condition.

Reference: Medical Terminology Root Analysis

Common Error: Confusing "-pathy" (disease) with "-itis" (inflammation).

Question 305 (Multiple Choice) What does "bilateral" mean?

A) One side only B) Both sides C) Multiple sides D) No specific side

Answer: B.

"Bilateral" means affecting both sides, as in bilateral knee arthritis affecting both knees.

Reference: Medical Terminology Directional Terms

Common Error: Confusing bilateral with unilateral (one side) or multilateral.

Question 306 (Scenario-Based) A patient has "diabetic retinopathy." This condition affects which body system?

A) Endocrine system B) Visual system C) Nervous system D) Both endocrine and visual systems

Answer: D.

Diabetic retinopathy is a complication of diabetes (endocrine) affecting the retina (visual system), demonstrating how conditions can involve multiple systems.

Reference: Multi-system Disease Terminology

Common Error: Choosing only one system when the condition involves complications affecting multiple systems.

Question 307 (Multiple Choice) The term "chronic" means:

A) Sudden onset B) Short duration C) Long-lasting or persistent D) Severe intensity

Answer: C.

"Chronic" refers to conditions that persist over a long period, typically months to years, as opposed to acute conditions.

Reference: Medical Terminology Temporal Descriptors

Common Error: Confusing chronic with acute or associating it with severity rather than duration.

Question 308 (Select All That Apply) Which prefixes indicate abnormal decrease? (Select all that apply)

A) Hypo- B) Hyper- C) Brady- D) Tachy- E) Oligo-

Answer: A, C, E.

"Hypo-" (below normal), "brady-" (slow), and "oligo-" (few/little) indicate decrease. "Hyper-" and "tachy-" indicate increase.

Reference: Medical Terminology Quantity/Rate Prefixes

Common Error: Including prefixes that indicate increase rather than decrease.

Question 309 (Multiple Choice) The root "cardi-" refers to:

A) Brain B) Heart C) Liver D) Kidney

Answer: B.

"Cardi-" or "cardio-" refers to the heart, as in cardiology (heart study) or cardiomyopathy (heart muscle disease).

Reference: Medical Terminology Anatomical Roots

Common Error: Confusing similar-sounding roots or mixing up organ-specific terminology.

Question 310 (Multiple Choice) What does "acute" mean in medical terminology?

A) Severe B) Mild C) Sudden onset or short duration D) Chronic condition

Answer: C.

"Acute" refers to conditions with sudden onset or short duration, as opposed to chronic conditions that persist over time.

Reference: Medical Terminology Temporal Classifications

Common Error: Associating acute with severity rather than timing and onset characteristics.

Question 311 (Multiple Choice) The suffix "-ectomy" means:

A) Study of B) Surgical removal C) Condition of D) Inflammation

Answer: B.

"-ectomy" means surgical removal, as in appendectomy (appendix removal) or cholecystectomy (gallbladder removal).

Reference: Medical Terminology Surgical Suffixes

Common Error: Confusing surgical suffixes like "-ectomy" (removal) with "-otomy" (cutting into).

Question 312 (Scenario-Based) A physician documents "tachycardia." This means the patient has:

A) Slow heart rate B) Fast heart rate C) Irregular heart rate D) Normal heart rate

Answer: B.

"Tachycardia" means fast heart rate, with "tachy-" meaning fast and "cardia" referring to heart rate.

Reference: Medical Terminology Cardiovascular Terms

Common Error: Confusing "tachy-" (fast) with "brady-" (slow).

Question 313 (Multiple Choice) The term "dyspnea" refers to:

A) Difficulty breathing B) Fast breathing C) Slow breathing D) Normal breathing

Answer: A.

"Dyspnea" means difficulty breathing, with "dys-" indicating difficulty or abnormal and "pnea" referring to breathing.

Reference: Medical Terminology Respiratory Terms

Common Error: Confusing dyspnea with other breathing-related terms like tachypnea (fast breathing).

Question 314 (Select All That Apply) Which terms indicate surgical procedures? (Select all that apply)

A) -ectomy B) -otomy C) -itis D) -ostomy E) -pathy

Answer: A, B, D.

"-ectomy" (removal), "-otomy" (cutting into), and "-ostomy" (creating an opening) are surgical suffixes. "-itis" means inflammation and "-pathy" means disease.

Reference: Medical Terminology Surgical Procedure Suffixes

Common Error: Including disease or condition suffixes rather than surgical procedure suffixes.

Question 315 (Multiple Choice) The prefix "poly-" means:

A) One B) Two C) Many D) None

Answer: C.

"Poly-" means many or multiple, as in polyuria (excessive urination) or polyneuropathy (multiple nerve damage).

Reference: Medical Terminology Quantity Prefixes

Common Error: Confusing numerical prefixes like mono- (one), bi- (two), and poly- (many).

Question 316 (Multiple Choice) What does "bradycardia" mean?

A) Fast heart rate B) Slow heart rate C) Irregular heart rate D) Absent heart rate

Answer: B.

"Bradycardia" means slow heart rate, with "brady-" meaning slow and "cardia" referring to heart rate.

Reference: Medical Terminology Cardiovascular Rate Terms

Common Error: Confusing "brady-" (slow) with "tachy-" (fast).

Question 317 (Multiple Choice) The root "pneumo-" refers to:

A) Blood B) Lung C) Heart D) Brain

Answer: B.

"Pneumo-" refers to lung or air, as in pneumonia (lung infection) or pneumothorax (air in chest cavity).

Reference: Medical Terminology Respiratory System Roots

Common Error: Confusing "pneumo-" with other organ-specific roots like "cardio-" (heart) or "hemo-" (blood).

Question 318 (Scenario-Based) A patient has "polyuria." This symptom involves:

A) Decreased urination B) Painful urination C) Excessive urination D) Blood in urine

Answer: C.

"Polyuria" means excessive urination, with "poly-" meaning many/excessive and "-uria" referring to urination.

Reference: Medical Terminology Urinary System Terms

Common Error: Confusing "poly-" (excessive) with "oligo-" (decreased) or other urinary prefixes.

Question 319 (Multiple Choice) The suffix "-megaly" indicates:

A) Inflammation B) Enlargement C) Surgical removal D) Pain

Answer: B.

"-megaly" means enlargement, as in cardiomegaly (enlarged heart) or hepatomegaly (enlarged liver).

Reference: Medical Terminology Size-Related Suffixes

Common Error: Confusing "-megaly" (enlargement) with "-itis" (inflammation) or other condition suffixes.

Question 320 (Multiple Choice) What does "hemiplegia" mean?

A) Complete paralysis B) Paralysis of one side of the body C) Paralysis of lower body D) Partial paralysis

Answer: B.

"Hemiplegia" means paralysis of one side of the body, with "hemi-" meaning half and "-plegia" meaning paralysis.

Reference: Medical Terminology Neurological Terms

Common Error: Confusing "hemi-" (half/one side) with other paralysis terms like "para-" (lower) or "quadri-" (all four limbs).

Question 321 (Multiple Choice) The term "anemia" refers to:

A) Excess red blood cells B) Deficiency of red blood cells or hemoglobin C) Abnormal white blood cells D) Blood clotting disorder

Answer: B.

"Anemia" refers to deficiency of red blood cells or hemoglobin, with "an-" meaning without and "-emia" referring to blood condition.

Reference: Medical Terminology Hematological Terms

Common Error: Confusing anemia (deficiency) with polycythemia (excess red blood cells).

Intermediate Level

Question 322 (Multiple Choice) The term "myocardial infarction" literally means:

A) Heart muscle inflammation B) Heart muscle death due to lack of blood supply C) Heart muscle enlargement D) Heart muscle weakness

Answer: B.

"Myocardial infarction" means death of heart muscle tissue due to lack of blood supply, with "myo-" (muscle), "cardial" (heart), and "infarction" (tissue death from lack of blood).

Reference: Medical Terminology Complex Cardiovascular Terms

Common Error: Confusing infarction (tissue death) with other cardiac conditions like inflammation or dysfunction.

Question 323 (Scenario-Based) A patient has "diabetic nephropathy with proteinuria." This indicates:

A) Diabetes with kidney disease and protein in urine B) Diabetes with nerve damage and protein deficiency C) Kidney infection with diabetes D) Protein diabetes with kidney stones

Answer: A.

Diabetic nephropathy is kidney disease caused by diabetes, and proteinuria is protein in the urine, indicating kidney damage.

Reference: Medical Terminology Diabetic Complications

Common Error: Confusing nephropathy (kidney disease) with neuropathy (nerve disease) or misunderstanding proteinuria.

Question 324 (Multiple Choice) What does "atherosclerosis" mean?

A) Hardening of arteries due to calcium deposits B) Hardening of arteries due to fatty plaque buildup C) Inflammation of arteries D) Narrowing of arteries due to spasm

Answer: B.

"Atherosclerosis" is hardening of arteries due to fatty plaque buildup, with "athero-" (fatty plaque) and "sclerosis" (hardening).

Reference: Medical Terminology Vascular Disease Terms

Common Error: Confusing atherosclerosis with arteriosclerosis (general artery hardening) or arteritis (artery inflammation).

Question 325 (Select All That Apply) Which terms indicate decreased function or deficiency? (Select all that apply)

A) Hypothyroidism B) Hyperthyroidism C) Bradycardia D) Tachycardia E) Oliguria

Answer: A, C, E.

Hypothyroidism (decreased thyroid function), bradycardia (slow heart rate), and oliguria (decreased urination) indicate decreased function. The others indicate increased function.

Reference: Medical Terminology Function-Related Terms

Common Error: Including terms that indicate increased rather than decreased function.

Question 326 (Multiple Choice) The term "cardiomyopathy" means:

A) Heart muscle inflammation B) Heart muscle disease C) Heart muscle enlargement D) Heart muscle death

Answer: B.

"Cardiomyopathy" means heart muscle disease, with "cardio-" (heart), "myo-" (muscle), and "-pathy" (disease).

Reference: Medical Terminology Cardiac Disease Terms

Common Error: Confusing "-pathy" (disease) with "-itis" (inflammation) or other specific pathological processes.

Question 327 (Multiple Choice) What does "peripheral arterial disease" refer to?

A) Disease of arteries in the extremities B) Disease of central arteries near the heart C) Disease of all arteries in the body D) Disease of arteries and veins together

Answer: A.

Peripheral arterial disease (PAD) refers to disease of arteries in the extremities, typically legs, away from the central circulation.

Reference: Medical Terminology Vascular System Locations

Common Error: Confusing peripheral (extremities) with central or coronary (heart) circulation.

Question 328 (Scenario-Based) A patient has "chronic obstructive pulmonary disease with acute exacerbation." This means:

A) Long-term lung disease with sudden worsening B) Short-term lung disease with gradual improvement C) Lung infection with breathing difficulty D) Allergic lung disease with seasonal flare

Answer: A.

COPD is a chronic (long-term) lung disease involving airway obstruction, and acute exacerbation means sudden worsening of symptoms.

Reference: Medical Terminology Respiratory Disease Progression

Common Error: Confusing chronic vs. acute or not understanding exacerbation as worsening.

Question 329 (Multiple Choice) The term "cerebrovascular accident" refers to:

A) Head injury from trauma B) Brain tumor C) Stroke D) Brain infection

Answer: C.

Cerebrovascular accident (CVA) is another term for stroke, involving disruption of blood supply to the brain.

Reference: Medical Terminology Neurological Events

Common Error: Confusing CVA with other brain conditions like trauma or infection.

Question 330 (Multiple Choice) What does "diabetic retinopathy" indicate?

A) Diabetes affecting the kidneys B) Diabetes affecting the eyes C) Diabetes affecting the nerves D) Diabetes affecting the heart

Answer: B.

Diabetic retinopathy is diabetes affecting the retina in the eyes, with "retino-" referring to the retina and "-pathy" meaning disease.

Reference: Medical Terminology Diabetic Complications by System

Common Error: Confusing retinopathy (eye) with nephropathy (kidney) or neuropathy (nerve).

Question 331 (Select All That Apply) Which terms relate to blood disorders? (Select all that apply)

A) Anemia B) Leukemia C) Pneumonia D) Thrombocytopenia E) Hepatitis

Answer: A, B, D.

Anemia (red blood cell deficiency), leukemia (white blood cell cancer), and thrombocytopenia (platelet deficiency) are blood disorders. Pneumonia affects lungs, hepatitis affects liver.

Reference: Medical Terminology Hematological System

Common Error: Including conditions that affect other systems despite potential blood-related complications.

Question 332 (Multiple Choice) The term "congestive heart failure" indicates:

A) Heart attack B) Heart rhythm abnormality C) Heart's inability to pump blood effectively D) Heart valve disease

Answer: C.

Congestive heart failure means the heart cannot pump blood effectively, leading to fluid buildup (congestion) in tissues.

Reference: Medical Terminology Heart Function Disorders

Common Error: Confusing heart failure with heart attack (myocardial infarction) or rhythm disorders.

Question 333 (Multiple Choice) What does "hypertensive nephrosclerosis" mean?

A) High blood pressure causing kidney scarring B) Kidney disease causing high blood pressure C) Kidney stones with high blood pressure D) Kidney infection with blood pressure changes

Answer: A.

Hypertensive nephrosclerosis is kidney scarring (sclerosis) caused by high blood pressure (hypertension).

Reference: Medical Terminology Hypertensive Complications

Common Error: Confusing cause and effect between hypertension and kidney disease.

Question 334 (Scenario-Based) A patient has "rheumatoid arthritis with systemic manifestations." This means:

A) Joint inflammation affecting only hands and feet B) Joint inflammation affecting the entire body C) Joint inflammation with effects on other organ systems D) Infectious joint inflammation

Answer: C.

Rheumatoid arthritis with systemic manifestations means the autoimmune joint inflammation also affects other organ systems beyond joints.

Reference: Medical Terminology Systemic Disease Terminology

Common Error: Limiting systemic to meaning "whole body joints" rather than "affecting multiple organ systems."

Question 335 (Multiple Choice) The term "polyneuropathy" refers to:

A) Single nerve damage B) Multiple nerve damage C) Brain nerve damage D) Spinal cord nerve damage

Answer: B.

"Polyneuropathy" means damage to multiple nerves, with "poly-" (many) and "neuropathy" (nerve disease).

Reference: Medical Terminology Neurological System Terms

Common Error: Confusing "poly-" (multiple) with "mono-" (single) or limiting to specific nerve locations.

Question 336 (Multiple Choice) What does "acute respiratory failure" mean?

A) Long-term breathing problems B) Sudden inability of lungs to provide adequate oxygen C) Lung infection D) Allergic breathing reaction

Answer: B.

Acute respiratory failure is sudden inability of the respiratory system to maintain adequate oxygenation or carbon dioxide removal.

Reference: Medical Terminology Respiratory Emergency Terms

Common Error: Confusing acute respiratory failure with chronic conditions or specific lung diseases.

Question 337 (Select All That Apply) Which terms indicate inflammatory conditions? (Select all that apply)

A) Arthritis B) Hepatitis C) Diabetes D) Gastritis E) Hypertension

Answer: A, B, D.

Arthritis (joint inflammation), hepatitis (liver inflammation), and gastritis (stomach inflammation) are inflammatory conditions. Diabetes and hypertension are not primarily inflammatory.

Reference: Medical Terminology Inflammatory Conditions

Common Error: Including non-inflammatory diseases that may have inflammatory components.

Question 338 (Multiple Choice) The term "ischemic cardiomyopathy" means:

A) Heart muscle disease due to infection B) Heart muscle disease due to inadequate blood supply C) Heart muscle disease due to high blood pressure D) Heart muscle disease due to genetic factors

Answer: B.

Ischemic cardiomyopathy is heart muscle disease caused by inadequate blood supply (ischemia) to the heart muscle.

Reference: Medical Terminology Cardiac Etiology Terms

Common Error: Confusing ischemic (inadequate blood supply) with other causes of cardiomyopathy.

Question 339 (Multiple Choice) What does "diabetic gastroparesis" indicate?

A) Diabetes affecting stomach emptying B) Diabetes affecting stomach acid production C) Diabetes affecting stomach lining D) Diabetes affecting stomach size

Answer: A.

Diabetic gastroparesis is delayed stomach emptying due to diabetes affecting the nerves that control stomach muscles.

Reference: Medical Terminology Diabetic Gastrointestinal Complications

Common Error: Confusing gastroparesis (delayed emptying) with other stomach-related complications.

Question 340 (Scenario-Based) A patient has "end-stage renal disease requiring dialysis." This means:

A) Early kidney disease with medication treatment B) Moderate kidney disease with diet changes C) Severe kidney failure requiring artificial kidney function D) Kidney infection requiring antibiotics

Answer: C.

End-stage renal disease (ESRD) is severe kidney failure where the kidneys can no longer function adequately, requiring dialysis or transplant.

Reference: Medical Terminology Renal Disease Staging

Common Error: Not understanding "end-stage" as the most severe level of disease.

Question 341 (Multiple Choice) The term "atrial fibrillation" refers to:

A) Heart muscle weakness B) Irregular heart rhythm from upper chambers C) Heart valve disease D) Heart artery blockage

Answer: B.

Atrial fibrillation is an irregular heart rhythm originating from the atria (upper heart chambers) with rapid, chaotic electrical activity.

Reference: Medical Terminology Cardiac Rhythm Disorders

Common Error: Confusing atrial fibrillation with other cardiac conditions like heart failure or coronary disease.

Question 342 (Multiple Choice) What does "chronic kidney disease stage 4" indicate?

A) Mild kidney damage B) Moderate kidney damage C) Severe kidney damage D) Kidney failure requiring dialysis

Answer: C.

CKD stage 4 indicates severe kidney damage with significantly reduced kidney function (GFR 15-29 mL/min/1.73m^2).

Reference: Medical Terminology Chronic Kidney Disease Staging

Common Error: Not understanding the CKD staging system or confusing stage 4 with end-stage (stage 5).

Question 343 (Select All That Apply) Which terms indicate heart rhythm abnormalities? (Select all that apply)

A) Tachycardia B) Bradycardia C) Atrial fibrillation D) Heart failure E) Arrhythmia

Answer: A, B, C, E.

Tachycardia (fast rate), bradycardia (slow rate), atrial fibrillation (irregular rhythm), and arrhythmia (abnormal rhythm) are rhythm disorders. Heart failure is a pumping problem.

Reference: Medical Terminology Cardiac Rhythm Classifications

Common Error: Including heart failure, which is a pumping disorder rather than rhythm disorder.

Question 344 (Multiple Choice) The term "diabetic neuropathy" refers to:

A) Diabetes affecting the kidneys B) Diabetes affecting the nerves C) Diabetes affecting the eyes D) Diabetes affecting the heart

Answer: B.

Diabetic neuropathy is nerve damage caused by diabetes, with "neuro-" referring to nerves and "-pathy" meaning disease.

Reference: Medical Terminology Diabetic Neurological Complications

Common Error: Confusing neuropathy (nerve) with nephropathy (kidney) or retinopathy (eye).

Question 345 (Multiple Choice) What does "hypertensive heart disease" mean?

A) Heart disease causing high blood pressure B) High blood pressure causing heart disease C) Heart infection with blood pressure changes D) Heart rhythm problems with hypertension

Answer: B.

Hypertensive heart disease is heart damage caused by chronic high blood pressure, including conditions like left ventricular hypertrophy.

Reference: Medical Terminology Hypertensive Cardiac Complications

Common Error: Confusing the causal relationship between hypertension and heart disease.

Question 346 (Scenario-Based) A patient has "systolic heart failure with reduced ejection fraction." This indicates:

A) Problem with heart filling B) Problem with heart pumping/squeezing C) Problem with heart rhythm D) Problem with heart valves

Answer: B.

Systolic heart failure involves problems with the heart's ability to pump/squeeze effectively, resulting in reduced ejection fraction.

Reference: Medical Terminology Heart Failure Classifications

Common Error: Confusing systolic (pumping) with diastolic (filling) heart failure.

Question 347 (Multiple Choice) The term "peripheral neuropathy" refers to:

A) Brain nerve damage B) Spinal cord nerve damage C) Nerve damage in extremities D) Central nervous system damage

Answer: C.

Peripheral neuropathy is nerve damage in the extremities (hands, feet, arms, legs) outside the central nervous system.

Reference: Medical Terminology Neurological System Locations

Common Error: Confusing peripheral (extremities) with central nervous system involvement.

Question 348 (Multiple Choice) What does "acute myocardial infarction" mean?

A) Chronic heart muscle weakness B) Sudden heart muscle death due to blocked artery C) Heart muscle inflammation D) Heart rhythm abnormality

Answer: B.

Acute myocardial infarction (heart attack) is sudden death of heart muscle due to blocked coronary artery cutting off blood supply.

Reference: Medical Terminology Acute Cardiac Events

Common Error: Confusing myocardial infarction with other cardiac conditions like myocarditis or heart failure.

Question 349 (Multiple Choice) The term "diabetic ketoacidosis" indicates:

A) Diabetes with low blood sugar B) Diabetes with high blood sugar and acid buildup C) Diabetes with kidney complications D) Diabetes with nerve complications

Answer: B.

Diabetic ketoacidosis (DKA) is a serious diabetes complication with very high blood sugar and dangerous acid buildup in the blood.

Reference: Medical Terminology Diabetic Emergency Conditions

Common Error: Confusing DKA with other diabetic complications or hypoglycemia.

Question 350 (Select All That Apply) Which terms indicate kidney-related conditions? (Select all that apply)

A) Nephritis B) Nephropathy C) Hepatitis D) Nephrosclerosis E) Carditis

Answer: A, B, D.

Nephritis (kidney inflammation), nephropathy (kidney disease), and nephrosclerosis (kidney scarring) are kidney-related. Hepatitis affects liver, carditis affects heart.

Reference: Medical Terminology Renal System Terms

Common Error: Including terms from other organ systems that sound similar or may have secondary kidney effects.

Question 351 (Multiple Choice) What does "chronic obstructive pulmonary disease" indicate?

A) Temporary lung infection B) Long-term lung disease with airway obstruction C) Lung cancer D) Allergic lung reaction

Answer: B.

COPD is a chronic (long-term) lung disease characterized by persistent airway obstruction, typically from emphysema and chronic bronchitis.

Reference: Medical Terminology Chronic Respiratory Diseases

Common Error: Confusing COPD with acute lung conditions, infections, or malignancies.

Question 352 (Multiple Choice) The term "cerebrovascular disease" refers to:

A) Brain tumor B) Brain infection C) Blood vessel disease affecting the brain D) Brain trauma

Answer: C.

Cerebrovascular disease involves blood vessel disorders affecting the brain, including stroke, aneurysms, and vascular malformations.

Reference: Medical Terminology Neurological Vascular Conditions

Common Error: Confusing cerebrovascular disease with other brain pathologies not involving blood vessels.

Question 353 (Scenario-Based) A patient has "rheumatoid arthritis with extraarticular manifestations." This means:

A) Arthritis affecting only large joints B) Arthritis affecting only small joints C) Arthritis with complications outside the joints D) Infectious arthritis

Answer: C.

Extraarticular manifestations are complications of rheumatoid arthritis affecting organs and systems outside the joints.

Reference: Medical Terminology Systemic Autoimmune Disease Terms

Common Error: Not understanding "extraarticular" as meaning "outside the joints."

Question 354 (Multiple Choice) What does "pulmonary edema" mean?

A) Lung infection B) Lung collapse C) Fluid accumulation in the lungs D) Lung scarring

Answer: C.

Pulmonary edema is abnormal accumulation of fluid in the lungs, often due to heart failure or other causes.

Reference: Medical Terminology Pulmonary Fluid Disorders

Common Error: Confusing pulmonary edema with other lung conditions like pneumonia or pneumothorax.

Question 355 (Multiple Choice) The term "hypertensive nephropathy" indicates:

A) Kidney disease causing high blood pressure B) High blood pressure causing kidney disease C) Kidney infection with blood pressure changes D) Kidney stones with hypertension

Answer: B.

Hypertensive nephropathy is kidney disease caused by chronic high blood pressure damaging kidney blood vessels.

Reference: Medical Terminology Hypertensive Organ Damage

Common Error: Reversing the causal relationship between hypertension and kidney disease.

Question 356 (Multiple Choice) What does "diastolic heart failure" mean?

A) Problem with heart pumping B) Problem with heart filling C) Problem with heart rhythm D) Problem with heart valves

Answer: B.

Diastolic heart failure involves problems with the heart's ability to fill properly during the relaxation phase (diastole).

Reference: Medical Terminology Heart Failure Physiology

Common Error: Confusing diastolic (filling) with systolic (pumping) heart failure.

Question 357 (Select All That Apply) Which terms indicate liver-related conditions? (Select all that apply)

A) Hepatitis B) Nephritis C) Hepatomegaly D) Cirrhosis E) Gastritis

Answer: A, C, D.

Hepatitis (liver inflammation), hepatomegaly (liver enlargement), and cirrhosis (liver scarring) are liver-related. Nephritis affects kidneys, gastritis affects stomach.

Reference: Medical Terminology Hepatic System Terms

Common Error: Including conditions from other gastrointestinal or abdominal organs.

Question 358 (Multiple Choice) The term "diabetic foot ulcer" refers to:

A) Foot infection in diabetics B) Open wound on foot due to diabetic complications C) Foot swelling in diabetics D) Foot numbness in diabetics

Answer: B.

Diabetic foot ulcer is an open wound on the foot that develops due to diabetic complications like neuropathy and poor circulation.

Reference: Medical Terminology Diabetic Dermatological Complications

Common Error: Confusing ulcers (open wounds) with other foot complications like infections or neuropathy.

Question 359 (Multiple Choice) What does "coronary artery disease" indicate?

A) Disease of arteries in the extremities B) Disease of arteries supplying the heart muscle C) Disease of all arteries in the body D) Disease of heart valves

Answer: B.

Coronary artery disease (CAD) involves disease of the arteries that supply blood to the heart muscle itself.

Reference: Medical Terminology Cardiac Vascular Conditions

Common Error: Confusing coronary (heart) arteries with peripheral arteries or other cardiac structures.

Question 360 (Scenario-Based) A patient has "chronic kidney disease with anemia." This indicates:

A) Kidney disease with unrelated blood disorder B) Kidney disease causing decreased red blood cell production C) Blood disorder causing kidney disease D) Kidney infection with blood loss

Answer: B.

CKD commonly causes anemia because damaged kidneys produce less erythropoietin, the hormone that stimulates red blood cell production.

Reference: Medical Terminology Renal-Hematological Relationships

Common Error: Not understanding the causal relationship between kidney disease and anemia.

Question 361 (Multiple Choice) The term "ventricular tachycardia" refers to:

A) Fast heart rate from upper chambers B) Fast heart rate from lower chambers C) Slow heart rate from any chamber D) Irregular heart rate from any chamber

Answer: B.

Ventricular tachycardia is a fast heart rhythm originating from the ventricles (lower heart chambers).

Reference: Medical Terminology Specific Cardiac Arrhythmias

Common Error: Confusing ventricular (lower chambers) with atrial (upper chambers) arrhythmias.

Question 362 (Multiple Choice) What does "septic shock" mean?

A) Severe infection with dangerously low blood pressure B) Heart attack with shock C) Allergic reaction with shock D) Blood loss with shock

Answer: A.

Septic shock is a life-threatening condition where severe infection causes dangerous drops in blood pressure and organ dysfunction.

Reference: Medical Terminology Infectious Emergency Conditions

Common Error: Confusing septic shock with other types of shock or less severe infections.

Question 363 (Multiple Choice) The term "morbid obesity" indicates:

A) Slightly overweight B) Moderately overweight C) Severely obese with health risks D) Normal weight with health issues

Answer: C.

Morbid obesity is severe obesity (typically BMI ≥40 or ≥35 with comorbidities) that significantly increases health risks and mortality.

Reference: Medical Terminology Obesity Classifications

Common Error: Not understanding the severity implied by "morbid" in medical terminology.

Question 364 (Multiple Choice) What does "acute respiratory distress syndrome" (ARDS) mean?

A) Chronic breathing difficulty B) Sudden severe lung inflammation and failure C) Allergic breathing reaction D) Lung infection

Answer: B.

ARDS is sudden, severe lung inflammation causing widespread damage and respiratory failure.

Reference: Medical Terminology Critical Care Respiratory Conditions

Common Error: Confusing ARDS with other respiratory conditions or chronic diseases.

Question 365 (Select All That Apply) Which terms indicate heart valve problems? (Select all that apply)

A) Mitral regurgitation B) Aortic stenosis C) Atrial fibrillation D) Tricuspid insufficiency E) Myocardial infarction

Answer: A, B, D.

Mitral regurgitation (valve leakage), aortic stenosis (valve narrowing), and tricuspid insufficiency (valve leakage) are valve problems. Atrial fibrillation is rhythm disorder, MI is muscle damage.

Reference: Medical Terminology Cardiac Valve Pathology

Common Error: Including rhythm disorders or muscle damage rather than specific valve pathology.

Question 366 (Multiple Choice) The term "diabetic coma" refers to:

A) Sleep disorder in diabetics B) Unconsciousness due to diabetes complications C) Confusion in diabetics D) Memory loss in diabetics

Answer: B.

Diabetic coma is unconsciousness resulting from severe diabetes complications like very high or very low blood sugar.

Reference: Medical Terminology Diabetic Emergency Neurological Complications

Common Error: Confusing diabetic coma with less severe neurological symptoms of diabetes.

Question 367 (Multiple Choice) What does "hypertensive crisis" indicate?

A) Mild blood pressure elevation B) Moderately high blood pressure C) Severely high blood pressure requiring immediate treatment D) Low blood pressure emergency

Answer: C.

Hypertensive crisis is severely elevated blood pressure (typically >180/120) that requires immediate medical treatment to prevent organ damage.

Reference: Medical Terminology Hypertensive Emergency Classifications

Common Error: Not understanding the urgency and severity implied by "crisis" in medical terminology.

Question 368 (Scenario-Based) A patient has "acute decompensated heart failure." This means:

A) New onset heart failure B) Sudden worsening of chronic heart failure C) Heart failure that has improved D) Heart failure with normal symptoms

Answer: B.

Acute decompensated heart failure is sudden worsening of previously stable chronic heart failure, often requiring hospitalization.

Reference: Medical Terminology Heart Failure Exacerbation Terms

Common Error: Confusing decompensation (worsening) with new onset or improvement.

Question 369 (Multiple Choice) The term "pulmonary embolism" refers to:

A) Lung infection B) Blood clot blocking lung arteries C) Lung collapse D) Fluid in lungs

Answer: B.

Pulmonary embolism is blockage of lung arteries by blood clots, typically originating from leg veins.

Reference: Medical Terminology Pulmonary Vascular Emergencies

Common Error: Confusing pulmonary embolism with other lung conditions like pneumonia or pneumothorax.

Question 370 (Multiple Choice) What does "diabetic autonomic neuropathy" indicate?

A) Diabetes affecting sensory nerves B) Diabetes affecting motor nerves C) Diabetes affecting nerves controlling automatic body functions D) Diabetes affecting brain nerves

Answer: C.

Diabetic autonomic neuropathy affects nerves controlling automatic functions like heart rate, digestion, and blood pressure regulation.

Reference: Medical Terminology Diabetic Neurological System Classifications

Common Error: Confusing autonomic (automatic functions) with sensory or motor nerve types.

Question 371 (Multiple Choice) The term "systemic lupus erythematosus" indicates:

A) Localized skin inflammation B) Autoimmune disease affecting multiple organ systems C) Bacterial infection D) Joint disease only

Answer: B.

Systemic lupus erythematosus (SLE) is an autoimmune disease that can affect multiple organ systems throughout the body.

Reference: Medical Terminology Systemic Autoimmune Diseases

Common Error: Limiting lupus to skin or joint involvement rather than understanding its systemic nature.

Question 372 (Multiple Choice) What does "chronic renal insufficiency" mean?

A) Temporary kidney problems B) Acute kidney failure C) Long-term decreased kidney function D) Kidney infection

Answer: C.

Chronic renal insufficiency is long-term decreased kidney function, similar to chronic kidney disease.

Reference: Medical Terminology Chronic Renal Conditions

Common Error: Confusing chronic (long-term) with acute (sudden) kidney conditions.

Question 373 (Multiple Choice) The term "unstable angina" refers to:

A) Predictable chest pain with exertion B) Unpredictable chest pain that may occur at rest C) Heart attack D) Heart rhythm problem

Answer: B.

Unstable angina is unpredictable chest pain due to coronary artery disease that may occur at rest and indicates high risk for heart attack.

Reference: Medical Terminology Acute Coronary Syndrome Classifications

Common Error: Confusing unstable angina with stable angina or myocardial infarction.

Advanced Level

Question 374 (Multiple Choice) The pathophysiological term "diabetic microangiopathy" refers to:

A) Large blood vessel complications of diabetes B) Small blood vessel complications of diabetes C) Nerve complications of diabetes D) Kidney complications of diabetes

Answer: B.

Diabetic microangiopathy refers to small blood vessel (capillary) damage caused by diabetes, affecting organs like kidneys, eyes, and nerves.

Reference: Medical Terminology Advanced Diabetic Vascular Pathology

Common Error: Confusing microangiopathy (small vessels) with macroangiopathy (large vessels).

Question 375 (Scenario-Based) A patient has "ischemic cardiomyopathy with reduced ejection fraction and functional mitral regurgitation." This complex condition involves:

A) Heart muscle disease due to poor blood supply with pumping problems and secondary valve leakage B) Heart infection with valve damage and rhythm problems C) Congenital heart disease with multiple defects D) Heart muscle inflammation with primary valve disease

Answer: A.

This describes heart muscle disease from inadequate blood supply, resulting in poor pumping function and secondary mitral valve leakage due to heart shape changes.

Reference: Medical Terminology Complex Cardiac Pathophysiology

Common Error: Not understanding the interconnected nature of ischemic cardiomyopathy, systolic dysfunction, and secondary valve problems.

Question 376 (Multiple Choice) The term "diabetic nephrosclerosis with proteinuria and hypertension" indicates:

A) Simple kidney infection in a diabetic B) Complex diabetic kidney disease with scarring, protein loss, and high blood pressure C) Kidney stones with diabetes D) Temporary kidney problems

Answer: B.

This describes advanced diabetic kidney disease involving scarring (sclerosis), protein leakage (proteinuria), and associated hypertension.

Reference: Medical Terminology Advanced Diabetic Nephropathy

Common Error: Oversimplifying the complex pathophysiology of advanced diabetic kidney disease.

Question 377 (Multiple Choice) What does "acute-on-chronic systolic heart failure with pulmonary edema" mean?

A) New heart failure with lung problems B) Sudden worsening of long-term heart pumping problems causing fluid in lungs C) Heart rhythm problem with breathing difficulty D) Heart infection with lung complications

Answer: B.

This describes sudden deterioration of chronic heart pumping problems, resulting in fluid backup into the lungs.

Reference: Medical Terminology Complex Heart Failure Presentations

Common Error: Not understanding the acute-on-chronic concept or the relationship between heart failure and pulmonary edema.

Question 378 (Select All That Apply) Advanced diabetic complications terminology includes: (Select all that apply)

A) Proliferative diabetic retinopathy B) Diabetic gastroparesis C) Diabetic ketoacidosis D) Diabetic dermopathy E) Diabetic mastopathy

Answer: A, B, C, D, E.

All listed terms represent specific advanced diabetic complications affecting different organ systems: eyes, gastrointestinal, metabolic, skin, and breast tissue.

Reference: Medical Terminology Comprehensive Diabetic Complications

Common Error: Not recognizing the full spectrum of diabetic complications across multiple organ systems.

Question 379 (Multiple Choice) The term "hypertensive encephalopathy" refers to:

A) High blood pressure causing headaches B) Brain damage from severely elevated blood pressure C) Stroke with high blood pressure D) Memory problems in hypertensive patients

Answer: B.

Hypertensive encephalopathy is brain dysfunction caused by severely elevated blood pressure, a medical emergency requiring immediate treatment.

Reference: Medical Terminology Hypertensive Neurological Emergencies

Common Error: Confusing with less severe hypertensive symptoms or other neurological conditions.

Question 380 (Scenario-Based) A patient has "chronic kidney disease stage 4 with secondary hyperparathyroidism and renal osteodystrophy." This indicates:

A) Simple kidney disease with bone problems B) Complex kidney disease causing hormone imbalances and bone disease C) Kidney infection with calcium problems D) Temporary kidney problems with joint pain

Answer: B.

Advanced CKD causes complex mineral and hormone imbalances, leading to parathyroid hormone elevation and bone disease.

Reference: Medical Terminology CKD Mineral and Bone Disorders

Common Error: Not understanding the complex systemic effects of advanced chronic kidney disease.

Question 381 (Multiple Choice) What does "diabetic autonomic gastroparesis with delayed gastric emptying" indicate?

A) Stomach infection in diabetics B) Diabetes affecting stomach nerves causing slow stomach emptying C) Diabetic diet problems D) Stomach ulcers in diabetics

Answer: B.

This describes diabetes damaging nerves that control stomach muscles, leading to delayed food movement out of the stomach.

Reference: Medical Terminology Advanced Diabetic Gastrointestinal Neuropathy

Common Error: Confusing gastroparesis with other gastrointestinal complications or infections.

Question 382 (Multiple Choice) The complex term "ischemic dilated cardiomyopathy with biventricular failure" means:

A) Heart muscle disease from poor blood supply affecting both heart pumping chambers B) Heart infection affecting multiple chambers C) Congenital heart defect with multiple problems D) Heart rhythm disorder affecting heart shape

Answer: A.

This describes heart muscle disease from inadequate blood supply, causing enlargement and failure of both ventricles (pumping chambers).

Reference: Medical Terminology Advanced Cardiac Pathophysiology

Common Error: Not understanding the progression from ischemia to dilated cardiomyopathy to biventricular failure.

Question 383 (Multiple Choice) What does "rheumatoid arthritis with extraarticular manifestations including rheumatoid lung disease" indicate?

A) Joint arthritis with unrelated lung problems B) Autoimmune joint disease affecting lungs as part of systemic involvement C) Lung infection in arthritis patients D) Arthritis medication side effects on lungs

Answer: B.

This describes systemic rheumatoid arthritis affecting lungs as part of the autoimmune process, not just joint involvement.

Reference: Medical Terminology Systemic Autoimmune Disease Manifestations

Common Error: Not understanding how systemic autoimmune diseases can affect multiple organ systems.

Question 384 (Scenario-Based) A patient has "diabetic proliferative retinopathy with macular edema and vitreous hemorrhage." This complex eye condition involves:

A) Simple diabetic eye problems B) Advanced diabetic eye disease with new blood vessel growth, fluid accumulation, and bleeding C) Eye infection in diabetics D) Eye pressure problems

Answer: B.

This describes advanced diabetic eye disease with abnormal blood vessel growth, fluid in the central vision area, and bleeding into the eye.

Reference: Medical Terminology Advanced Diabetic Ophthalmology

Common Error: Not understanding the complexity and severity of advanced diabetic retinopathy.

Question 385 (Multiple Choice) The term "acute decompensated heart failure with cardiorenal syndrome" indicates:

A) Heart failure with unrelated kidney problems B) Heart failure causing kidney dysfunction or vice versa C) Heart and kidney infections D) Medication side effects on heart and kidneys

Answer: B.

Cardiorenal syndrome describes the complex interaction where heart failure worsens kidney function or kidney problems worsen heart failure.

Reference: Medical Terminology Cardiorenal Pathophysiology

Common Error: Not understanding the bidirectional relationship between heart and kidney dysfunction.

Question 386 (Select All That Apply) Advanced cardiovascular terminology includes: (Select all that apply)

A) Non-ST elevation myocardial infarction (NSTEMI) B) Acute coronary syndrome C) Takotsubo cardiomyopathy D) Hypertrophic obstructive cardiomyopathy E) Arrhythmogenic right ventricular cardiomyopathy

Answer: A, B, C, D, E.

All represent advanced cardiovascular terminology describing specific types of heart attacks, acute coronary conditions, and specialized cardiomyopathies.

Reference: Medical Terminology Advanced Cardiovascular Pathology

Common Error: Not recognizing the specificity and clinical significance of advanced cardiovascular diagnostic terminology.

Question 387 (Multiple Choice) What does "diabetic chronic kidney disease with proteinuria and hypertensive nephrosclerosis" indicate?

A) Simple kidney problems in diabetes B) Complex kidney disease from multiple causes: diabetes and high blood pressure C) Kidney infection with diabetes D) Temporary kidney function changes

Answer: B.

This describes kidney disease from both diabetes and chronic high blood pressure, with protein leakage indicating significant damage.

Reference: Medical Terminology Multifactorial Renal Disease

Common Error: Not recognizing that chronic kidney disease often has multiple contributing factors.

Question 388 (Multiple Choice) The term "systolic and diastolic heart failure with preserved vs. reduced ejection fraction" distinguishes:

A) Different heart chambers involved B) Different phases of heart pumping cycle affected C) Different severity levels D) Different types of heart rhythm problems

Answer: B.

This terminology distinguishes heart failure affecting the pumping phase (systolic) vs. filling phase (diastolic), with different ejection fraction measurements.

Reference: Medical Terminology Advanced Heart Failure Classifications

Common Error: Confusing ejection fraction types with severity levels or anatomical differences.

Question 389 (Scenario-Based) A patient has "chronic obstructive pulmonary disease with cor pulmonale and pulmonary hypertension." This indicates:

A) Lung disease with unrelated heart problems B) Lung disease causing right heart failure due to lung blood pressure elevation C) Heart disease causing lung problems D) Infection affecting both lungs and heart

Answer: B.

COPD can cause pulmonary hypertension (high lung blood pressure), leading to cor pulmonale (right heart failure from lung disease).

Reference: Medical Terminology Cardiopulmonary Pathophysiology

Common Error: Not understanding how chronic lung disease can cause specific heart complications.

Question 390 (Multiple Choice) What does "diabetic polyneuropathy with autonomic involvement" mean?

A) Diabetes affecting single nerves B) Diabetes affecting multiple nerves including those controlling automatic body functions C) Diabetes affecting only sensory nerves D) Diabetes affecting only motor nerves

Answer: B.

This describes diabetes damaging multiple nerves, including both peripheral nerves and autonomic nerves controlling automatic functions.

Reference: Medical Terminology Comprehensive Diabetic Neuropathy

Common Error: Not understanding that diabetic neuropathy can affect different types of nerves simultaneously.

Question 391 (Multiple Choice) The complex term "hypertensive heart disease with left ventricular hypertrophy and diastolic dysfunction" indicates:

A) Simple high blood pressure B) High blood pressure causing heart muscle thickening and filling problems C) Heart rhythm problems with blood pressure D) Heart valve disease with hypertension

Answer: B.

This describes how chronic hypertension causes the left ventricle to thicken (hypertrophy) and develop filling problems (diastolic dysfunction).

Reference: Medical Terminology Hypertensive Cardiac Remodeling

Common Error: Not understanding the progressive cardiac changes caused by chronic hypertension.

Question 392 (Select All That Apply) Advanced renal pathophysiology terminology includes: (Select all that apply)

A) Acute tubular necrosis B) Chronic tubulointerstitial nephritis C) Diabetic glomerulosclerosis D) Hypertensive nephrosclerosis E) Polycystic kidney disease

Answer: A, B, C, D, E.

All represent specific advanced kidney pathology terms describing different mechanisms and locations of kidney damage.

Reference: Medical Terminology Advanced Nephrology Pathophysiology

Common Error: Not recognizing the anatomical and pathophysiological specificity of advanced renal terminology.

Question 393 (Multiple Choice) What does "acute respiratory failure with hypoxemic and hypercapnic components" indicate?

A) Simple breathing difficulty B) Lung failure with both low oxygen and high carbon dioxide retention C) Lung infection D) Allergic breathing reaction

Answer: B.

This describes complete respiratory failure where lungs cannot maintain adequate oxygen levels or remove carbon dioxide effectively.

Reference: Medical Terminology Advanced Respiratory Failure Classifications

Common Error: Not understanding the dual gas exchange failure in advanced respiratory failure.

Question 394 (Scenario-Based) A patient has "end-stage renal disease with uremic encephalopathy and metabolic acidosis." This indicates:

A) Simple kidney problems B) Complete kidney failure causing brain dysfunction and blood acid buildup C) Kidney infection with confusion D) Temporary kidney problems

Answer: B.

ESRD with uremic toxins affecting brain function and acid buildup from inability to maintain acid-base balance.

Reference: Medical Terminology ESRD Systemic Complications

Common Error: Not understanding the systemic effects of complete kidney failure.

Question 395 (Multiple Choice) The term "ischemic stroke with hemorrhagic transformation" means:

A) Two separate strokes B) Stroke from blood clot that later develops bleeding C) Stroke from bleeding that stops D) Minor stroke with complications

Answer: B.

This describes a stroke initially caused by blocked blood flow that subsequently develops bleeding in the affected brain tissue.

Reference: Medical Terminology Complex Stroke Pathophysiology

Common Error: Not understanding how ischemic strokes can develop secondary hemorrhagic complications.

Question 396 (Multiple Choice) What does "diabetic ketoacidosis with severe dehydration and electrolyte imbalance" indicate?

A) Mild diabetes complications B) Life-threatening diabetes emergency with multiple metabolic derangements C) Simple high blood sugar D) Diabetes medication side effects

Answer: B.

DKA with dehydration and electrolyte imbalances represents a complex, life-threatening metabolic emergency requiring intensive treatment.

Reference: Medical Terminology Diabetic Emergency Pathophysiology

Common Error: Not appreciating the complexity and severity of DKA with systemic complications.

Question 397 (Multiple Choice) The complex term "chronic systolic heart failure with cardiac cachexia and renal insufficiency" indicates:

A) Simple heart problems B) Advanced heart failure causing muscle wasting and kidney dysfunction C) Heart infection with weight loss D) Heart medication side effects

Answer: B.

This describes advanced heart failure syndrome causing muscle and fat loss (cachexia) and secondary kidney problems.

Reference: Medical Terminology Advanced Heart Failure Syndrome

Common Error: Not understanding cardiac cachexia as a sign of advanced heart failure or the cardiorenal connection.

Question 398 (Select All That Apply) Advanced diabetic complications requiring complex terminology include: (Select all that apply)

A) Diabetic nephropathy with nephrotic syndrome B) Diabetic retinopathy with tractional retinal detachment C) Diabetic neuropathy with Charcot arthropathy D) Diabetic cardiomyopathy E) Diabetic foot with osteomyelitis

Answer: A, B, C, D, E.

All represent advanced diabetic complications with complex pathophysiology requiring specific medical terminology.

Reference: Medical Terminology Advanced Diabetic Complications

Common Error: Not recognizing the complexity and systemic nature of advanced diabetic complications.

Question 399 (Multiple Choice) What does "acute coronary syndrome with non-ST elevation myocardial infarction and troponin elevation" indicate?

A) Simple chest pain B) Heart attack without specific EKG changes but with heart muscle damage markers C) Heart rhythm problem D) Heart valve disease

Answer: B.

NSTEMI represents heart attack with biochemical evidence (troponin) but without characteristic EKG changes seen in STEMI.

Reference: Medical Terminology Acute Coronary Syndrome Classifications

Common Error: Not understanding the distinction between different types of heart attacks based on EKG and biomarker findings.

Question 400 (Multiple Choice) The term "multiorgan dysfunction syndrome secondary to septic shock" indicates:

A) Single organ infection B) Infection causing failure of multiple organ systems C) Multiple unrelated health problems D) Medication side effects

Answer: B.

This describes the cascade where severe infection (septic shock) leads to dysfunction of multiple organ systems simultaneously.

Reference: Medical Terminology Critical Care Pathophysiology

Common Error: Not understanding how severe sepsis can cause systemic organ dysfunction beyond the initial infection site.

Question 401 (Scenario-Based) A patient has "hypertensive emergency with acute kidney injury and encephalopathy." This complex presentation involves:

A) Simple high blood pressure B) Dangerously high blood pressure causing immediate kidney and brain damage C) High blood pressure with unrelated problems D) Medication reaction

Answer: B.

Hypertensive emergency causes immediate organ damage, in this case affecting both kidneys (acute injury) and brain (encephalopathy).

Reference: Medical Terminology Hypertensive Emergency Complications

Common Error: Not understanding that hypertensive emergency involves immediate organ damage, not just elevated blood pressure.

Question 402 (Multiple Choice) What does "chronic kidney disease with mineral and bone disorder and secondary hyperparathyroidism" indicate?

A) Simple kidney problems B) Kidney disease causing complex calcium, phosphorus, and hormone imbalances affecting bones C) Kidney stones with bone pain D) Kidney infection with calcium problems

Answer: B.

CKD-MBD represents the complex disruption of mineral metabolism and bone health that occurs in advanced kidney disease.

Reference: Medical Terminology CKD Systemic Complications

Common Error: Not understanding the complex systemic mineral and hormone effects of chronic kidney disease.

Question 403 (Multiple Choice) The advanced term "dilated cardiomyopathy with reduced ejection fraction and functional mitral regurgitation" describes:

A) Simple heart enlargement B) Heart muscle disease causing enlargement, poor pumping, and secondary valve leakage C) Heart infection D) Congenital heart defect

Answer: B.

This describes the progression where heart muscle disease causes chamber enlargement, reduced pumping function, and secondary valve problems.

Reference: Medical Terminology Advanced Cardiomyopathy Pathophysiology

Common Error: Not understanding how cardiomyopathy leads to both systolic dysfunction and secondary valve regurgitation.

Question 404 (Multiple Choice) What does "end-stage liver disease with hepatorenal syndrome and portal hypertension" indicate?

A) Simple liver problems B) Advanced liver failure causing kidney dysfunction and increased liver blood pressure C) Liver infection D) Liver medication side effects

Answer: B.

This describes advanced liver disease with kidney failure (hepatorenal syndrome) and increased pressure in liver blood vessels (portal hypertension).

Reference: Medical Terminology Advanced Hepatology Pathophysiology

Common Error: Not understanding the complex systemic effects of end-stage liver disease.

Question 405 (Multiple Choice) The comprehensive term "diabetes mellitus with multiple complications including nephropathy, retinopathy, neuropathy, and accelerated atherosclerosis" represents:

A) Simple diabetes B) Complex diabetes with damage to multiple organ systems C) Diabetes medication side effects D) Multiple unrelated conditions

Answer: B.

This describes advanced diabetes with complications affecting kidneys, eyes, nerves, and blood vessels throughout the body.

Reference: Medical Terminology Comprehensive Diabetic Complications

Common Error: Not appreciating the systemic nature and interconnectedness of advanced diabetic complications across multiple organ systems.

Hcc Categories & Hierarchies

Foundation Level

Question 406 (Multiple Choice) What does HCC stand for?

A) Health Care Categories B) Hierarchical Condition Categories C) Hospital Care Codes D) Health Condition Classifications

Answer: B.

HCC stands for Hierarchical Condition Categories, which are groups of related medical conditions used for risk adjustment payments in Medicare Advantage.

Reference: CMS-HCC Model Overview

Common Error: Confusing HCC with other healthcare acronyms or classification systems.

Question 407 (Multiple Choice) How many HCC categories are there in the CMS-HCC Model V28?

A) 85 B) 115 C) 189 D) 254

Answer: B.

The CMS-HCC Model V28 contains 115 HCC categories that generate risk adjustment payments.

Reference: CMS-HCC Model V28 Category List

Common Error: Using HCC counts from previous model versions or including non-payment categories.

Question 408 (Select All That Apply) Which conditions belong to the same HCC hierarchy? (Select all that apply)

A) Diabetes without complications B) Diabetes with acute complications C) Diabetes with chronic complications D) Hypertension E) Heart failure

Answer: A, B, C.

All diabetes conditions are part of the diabetes hierarchy, where more severe forms suppress less severe forms. Hypertension and heart failure are in different hierarchies.

Reference: CMS-HCC Model V28 Hierarchy Tables

Common Error: Including unrelated conditions or not understanding hierarchical relationships within condition families.

Question 409 (Multiple Choice) What happens when a patient has both diabetes without complications (HCC 35) and diabetes with complications (HCC 18)?

A) Both HCCs generate payments B) Only HCC 35 generates payment C) Only HCC 18 generates payment D) The payments are averaged

Answer: C.

In hierarchical relationships, the more severe condition (diabetes with complications) suppresses the less severe condition to prevent double payment.

Reference: CMS-HCC Model V28 Diabetes Hierarchy

Common Error: Thinking both related conditions can contribute payments or that hierarchy works in reverse.

Question 410 (Multiple Choice) Which HCC category typically has the highest payment value?

A) Diabetes without complications B) Hypertension C) Major depressive disorder D) End-stage renal disease

Answer: D.

End-stage renal disease typically has among the highest HCC payment values due to the extremely high costs associated with dialysis and related care.

Reference: CMS-HCC Model V28 Coefficient Values

Common Error: Assuming all chronic conditions have similar payment values without considering cost differences.

Question 411 (Scenario-Based) A patient has documented congestive heart failure. Which HCC category does this map to?

A) HCC 221 - Heart Failure B) HCC 196 - Ischemic Heart Disease C) HCC 227 - Cardio-Respiratory Failure D) HCC 222 - Cardiorespiratory Failure

Answer: A.

Congestive heart failure maps to HCC 221 (Heart Failure) in the CMS-HCC Model V28.

Reference: CMS-HCC Model V28 ICD-10-CM Mapping

Common Error: Confusing heart failure with other cardiac conditions or using outdated HCC mappings.

Question 412 (Multiple Choice) The purpose of HCC hierarchies is to:

A) Rank conditions by severity B) Prevent double payment for related conditions C) Organize conditions by body system D) Determine treatment protocols

Answer: B.

HCC hierarchies prevent double payment when a patient has multiple related conditions by allowing only the most severe condition in each hierarchy to generate payment.

Reference: CMS-HCC Model Hierarchy Principles

Common Error: Misunderstanding hierarchies as severity rankings rather than payment prevention mechanisms.

Question 413 (Multiple Choice) Which statement about HCC categories is correct?

A) All chronic conditions map to HCCs B) Only inpatient diagnoses can generate HCCs C) HCCs require annual documentation to generate payments D) HCC values are the same for all conditions

Answer: C.

HCCs require annual documentation during qualifying face-to-face encounters to continue generating risk adjustment payments.

Reference: CMS Risk Adjustment Annual Documentation Requirements

Common Error: Assuming HCCs continue indefinitely once documented or that all conditions map to HCCs.

Question 414 (Select All That Apply) Which factors affect HCC payment values? (Select all that apply)

A) Historical medical costs for the condition B) Frequency of the condition in Medicare population C) Geographic location of the patient D) Provider specialty treating the condition E) Expected future medical costs

Answer: A, E.

HCC payment values are based on historical and expected future medical costs associated with specific conditions. Geography affects payment separately, and provider specialty doesn't affect HCC values.

Reference: CMS-HCC Model Development Methodology

Common Error: Including factors like provider specialty or geography that affect payment rates but not HCC coefficients.

Question 415 (Multiple Choice) Chronic kidney disease stage 4 maps to which HCC?

A) HCC 326 - Chronic Kidney Disease Stage 4 B) HCC 327 - Chronic Kidney Disease Stage 5 C) HCC 328 - End Stage Renal Disease D) HCC 329 - Kidney Transplant Status

Answer: A.

Chronic kidney disease stage 4 maps to HCC 326 in the CMS-HCC Model V28.

Reference: CMS-HCC Model V28 Kidney Disease Categories

Common Error: Confusing different CKD stages or using incorrect HCC numbers.

Question 416 (Multiple Choice) Which condition would NOT typically map to an HCC?

A) Diabetes mellitus type 2 B) Hypertension C) Common cold D) Major depressive disorder

Answer: C.

Acute, self-limiting conditions like common cold do not map to HCCs. HCCs are reserved for chronic conditions or conditions requiring ongoing medical management.

Reference: CMS-HCC Model Condition Selection Criteria

Common Error: Thinking all medical conditions map to HCCs or including acute, minor conditions.

Question 417 (Scenario-Based) A patient has both rheumatoid arthritis and osteoarthritis documented. How does this affect HCC payments?

A) Only rheumatoid arthritis generates payment B) Only osteoarthritis generates payment C) Both conditions generate separate payments D) The conditions are averaged for payment

Answer: C.

Rheumatoid arthritis and osteoarthritis are different conditions that can both generate HCC payments as they are not in the same hierarchy.

Reference: CMS-HCC Model V28 Arthritis Categories

Common Error: Assuming all arthritis conditions are hierarchically related.

Question 418 (Multiple Choice) What does "interaction" mean in HCC terminology?

A) Conditions that cannot occur together B) Additional payment for specific combinations of conditions C) Conditions that require specialist care D) Conditions that affect the same body system

Answer: B.

HCC interactions provide additional payment when specific combinations of conditions create higher costs than individual conditions would predict.

Reference: CMS-HCC Model V28 Interaction Terms

Common Error: Confusing interactions with hierarchies or clinical contraindications.

Question 419 (Multiple Choice) Which HCC category includes major depressive disorder?

A) HCC 154 - Bipolar Disorders B) HCC 155 - Major Depressive and Bipolar Disorders C) HCC 156 - Psychosis D) HCC 157 - Anxiety Disorders

Answer: B.

Major depressive disorder maps to HCC 155 (Major Depressive and Bipolar Disorders) in Model V28.

Reference: CMS-HCC Model V28 Mental Health Categories

Common Error: Using outdated HCC numbers or confusing different mental health categories.

Question 420 (Select All That Apply) Which conditions are typically in hierarchical relationships? (Select all that apply)

A) Different stages of the same disease B) Complicated vs uncomplicated versions of conditions C) Acute vs chronic forms of the same condition D) Conditions affecting different body systems E) Related conditions with different severity levels

Answer: A, B, E.

Hierarchies typically involve different stages, complicated vs uncomplicated versions, and related conditions with different severity levels. Acute vs chronic and different body systems usually aren't hierarchical.

Reference: CMS-HCC Model Hierarchy Design Principles

Common Error: Including temporal (acute vs chronic) or anatomical relationships that aren't typically hierarchical.

Question 421 (Multiple Choice) The HCC for "diabetes with chronic complications" is:

A) HCC 35 - Diabetes without Complications B) HCC 18 - Diabetes with Chronic Complications C) HCC 19 - Diabetes with Acute Complications D) HCC 20 - Diabetes with Other Complications

Answer: B.

Diabetes with chronic complications maps to HCC 18 in the CMS-HCC Model V28.

Reference: CMS-HCC Model V28 Diabetes Categories

Common Error: Using incorrect HCC numbers or confusing different types of diabetic complications.

Question 422 (Multiple Choice) Which statement about HCC payments is correct?

A) All HCCs have the same payment value B) HCC payments are added to demographic payments C) HCC payments replace demographic payments D) HCC payments are only for inpatient care

Answer: B.

HCC payments are added to demographic risk adjustment payments to create the total Risk Adjustment Factor (RAF) score.

Reference: CMS Risk Adjustment Payment Methodology

Common Error: Thinking HCCs replace rather than supplement demographic payments or have uniform values.

Question 423 (Scenario-Based) A Medicare Advantage member has diabetes, heart failure, and chronic kidney disease. How many HCCs might this generate?

A) One HCC for all conditions B) Two HCCs maximum C) Three separate HCCs D) No HCCs unless complications are present

Answer: C.

Each condition can generate separate HCC payments as they affect different body systems and are not hierarchically related.

Reference: CMS-HCC Model Multi-System Conditions

Common Error: Assuming conditions must be related to generate multiple HCCs or that there's a limit on HCC counts.

Question 424 (Multiple Choice) What is the primary difference between HCC 18 and HCC 35 in diabetes?

A) Type 1 vs Type 2 diabetes B) Presence vs absence of complications C) Insulin vs non-insulin treated D) Acute vs chronic diabetes

Answer: B.

HCC 18 is for diabetes with chronic complications while HCC 35 is for diabetes without complications, regardless of diabetes type or treatment.

Reference: CMS-HCC Model V28 Diabetes Hierarchy

Common Error: Focusing on diabetes type or treatment rather than presence of complications.

Question 425 (Multiple Choice) Which factor does NOT affect HCC mapping?

A) Specificity of the ICD-10-CM code B) Presence of complications C) Patient's age D) Clinical documentation

Answer: C.

Patient age affects demographic risk adjustment but does not change which HCC a specific condition maps to.

Reference: CMS-HCC Model Mapping Methodology

Common Error: Thinking patient demographics affect HCC mapping rather than just demographic risk scores.

Question 426 (Multiple Choice) The HCC model is updated to:

A) Reflect current medical costs and practices B) Add new medical conditions C) Respond to provider feedback D) All of the above

Answer: D.

HCC model updates incorporate current cost data, new medical knowledge, emerging conditions, and stakeholder feedback.

Reference: CMS-HCC Model Development Process

Common Error: Limiting model updates to single factors rather than comprehensive evidence-based revisions.

Intermediate Level

Question 427 (Multiple Choice) In CMS-HCC Model V28, which diabetes HCC hierarchy relationship is correct?

A) HCC 18 suppresses HCC 35 B) HCC 35 suppresses HCC 18 C) HCC 18 and 35 can both contribute D) Only one diabetes HCC can ever be present

Answer: A.

HCC 18 (Diabetes with Chronic Complications) suppresses HCC 35 (Diabetes without Complications) in the hierarchy.

Reference: CMS-HCC Model V28 Diabetes Hierarchy Table

Common Error: Reversing the hierarchy direction or not understanding that more severe conditions suppress less severe ones.

Question 428 (Scenario-Based) A patient has documented rheumatoid arthritis (maps to HCC 157) and also develops chronic kidney disease stage 4 (maps to HCC 326). An interaction exists between these conditions. What is the total HCC contribution?

A) Only the higher-value HCC B) HCC 157 + HCC 326 C) HCC 157 + HCC 326 + Interaction value D) Average of the two HCC values

Answer: C.

Both HCCs contribute plus the additional interaction value, as interactions provide extra payment for specific combinations that increase costs.

Reference: CMS-HCC Model V28 Interaction Tables

Common Error: Not including interaction values or thinking interactions replace rather than supplement individual HCC values.

Question 429 (Multiple Choice) Which factor most significantly changed HCC values between Model V24 and V28?

A) Inflation adjustments B) Updated medical cost data and recalibration C) Changes in Medicare demographics D) Provider payment negotiations

Answer: B.

Model V28 incorporated updated medical cost data and recalibrated coefficients based on more recent Medicare claims and cost patterns.

Reference: CMS-HCC Model V28 Development Documentation

Common Error: Assuming changes are due to simple inflation rather than comprehensive data recalibration.

Question 430 (Select All That Apply) Which HCC categories were significantly modified in Model V28? (Select all that apply)

A) Diabetes categories were consolidated B) Mental health categories were expanded C) Cancer categories were reorganized D) Cardiovascular categories remained unchanged E) Substance abuse categories were added

Answer: A, B, C.

V28 consolidated diabetes HCCs, expanded mental health categories, and reorganized cancer categories. Cardiovascular and substance abuse categories also saw changes.

Reference: CMS-HCC Model V28 Category Changes Summary

Common Error: Not recognizing the comprehensive nature of V28 changes across multiple condition categories.

Question 431 (Calculation Exercise) A patient has the following HCCs with their coefficients:

- HCC 18 (Diabetes with complications): 0.318
- HCC 221 (Heart failure): 0.368
- HCC 326 (CKD stage 4): 0.237
- Diabetes-Heart failure interaction: 0.154

What is the total HCC contribution to RAF?

A) 0.923 B) 1.077 C) 0.368 D) 0.709

Answer: B.

Total = 0.318 + 0.368 + 0.237 + 0.154 = 1.077. All HCCs and interactions are additive.

Reference: CMS-HCC Model V28 Coefficient Calculation

Common Error: Not including all HCCs or the interaction term in the calculation.

Question 432 (Multiple Choice) The concept of "disease interactions" in HCC models recognizes that:

A) Some diseases cannot occur together B) Certain combinations of diseases cost more than the sum of individual diseases C) Diseases in the same body system interact D) Drug interactions affect medical costs

Answer: B.

Disease interactions acknowledge that certain combinations create synergistic cost effects where total costs exceed what individual conditions would predict.

Reference: CMS-HCC Model Interaction Theory

Common Error: Confusing disease interactions with clinical contraindications or pharmacological interactions.

Question 433 (Multiple Choice) Which HCC hierarchy relationship is most clinically logical?

A) Acute myocardial infarction suppresses old myocardial infarction B) Chronic kidney disease stage 5 suppresses stage 3 C) Type 1 diabetes suppresses Type 2 diabetes D) Systolic heart failure suppresses diastolic heart failure

Answer: B.

More advanced stages of the same disease (CKD stage 5 vs 3) logically suppress less advanced stages to prevent double payment for disease progression.

Reference: CMS-HCC Model Clinical Hierarchy Logic

Common Error: Not understanding that hierarchies follow clinical progression and severity rather than temporal or type differences.

Question 434 (Scenario-Based) A health plan notices their HCC 18 (Diabetes with Complications) capture rate increased from 25% to 45% of diabetic members after implementing systematic diabetic complication screening. What is the most likely explanation?

A) The plan is inappropriately coding complications B) The diabetic population became sicker C) Improved clinical detection and documentation of existing complications D) The HCC model changed

Answer: C.

Systematic screening often reveals previously undiagnosed or undocumented complications that were clinically present but not captured.

Reference: HCC Capture Rate Improvement Analysis

Common Error: Assuming increases in HCC capture always indicate inappropriate coding rather than improved clinical detection.

Question 435 (Multiple Choice) The most significant impact of transitioning from V24 to V28 HCC model was:

A) Overall payment increases for all conditions B) Redistribution of payments based on updated cost data C) Elimination of all hierarchies D) Addition of 50 new HCC categories

Answer: B.

V28 redistributed payments based on updated cost data, with some conditions seeing increases and others decreases based on current medical cost patterns.

Reference: CMS-HCC Model V28 Payment Impact Analysis

Common Error: Expecting uniform payment changes rather than redistribution based on actual cost data updates.

Question 436 (Select All That Apply) Which factors influence HCC coefficient values? (Select all that apply)

A) Average annual medical costs for the condition B) Frequency of the condition in Medicare population C) Complexity of medical management required D) Availability of effective treatments E) Regional cost variations

Answer: A, C.

HCC coefficients primarily reflect average medical costs and management complexity. Frequency affects normalization, and regional costs are adjusted separately.

Reference: CMS-HCC Model Coefficient Development

Common Error: Including frequency or regional factors that affect other aspects of payment but not base HCC coefficients.

Question 437 (Multiple Choice) Advanced cancer HCC categories in V28 distinguish between:

A) Different cancer locations only B) Cancer staging and treatment status C) Benign vs malignant tumors D) Primary vs secondary cancers

Answer: B.

V28 cancer HCCs reflect staging, treatment status, and severity rather than just anatomical location.

Reference: CMS-HCC Model V28 Cancer Category Redesign

Common Error: Thinking cancer HCCs are organized only by anatomical location rather than clinical severity and prognosis.

Question 438 (Multiple Choice) The rationale for HCC hierarchies is primarily:

A) Clinical - more severe conditions require different treatment B) Administrative - to simplify coding requirements C) Financial - to prevent overpayment for related conditions D) Quality - to ensure appropriate specialist referrals

Answer: C.

HCC hierarchies primarily exist to prevent overpayment when patients have multiple related conditions that could otherwise generate duplicate payments.

Reference: CMS-HCC Model Hierarchy Design Rationale

Common Error: Focusing on clinical or administrative rationales rather than the primary financial/payment integrity purpose.

Question 439 (Scenario-Based) A Medicare Advantage plan's data shows that members with diabetes + depression have 30% higher costs than diabetes alone, but no interaction exists in the HCC model. What should the plan's response be?

A) Request CMS add this interaction to the model B) Develop targeted care management for this combination C) Avoid documenting both conditions simultaneously D) Code these members as having diabetic complications

Answer: B.

Plans should develop care management for clinically meaningful high-risk combinations even when not explicitly recognized in payment models.

Reference: Population Health Management Beyond HCC Models

Common Error: Waiting for model recognition rather than proactively managing identified high-risk populations.

Question 440 (Multiple Choice) Which statement about HCC interactions is correct?

A) Interactions always increase total payments B) Interactions can be positive or negative C) All condition combinations have interactions D) Interactions replace individual HCC payments

Answer: B.

HCC interactions can be positive (additional payment) or negative (payment reduction) depending on whether the combination increases or decreases expected costs.

Reference: CMS-HCC Model Interaction Methodology

Common Error: Assuming all interactions are positive or that interactions always supplement rather than sometimes reduce payments.

Question 441 (Multiple Choice) The largest HCC categories by Medicare population prevalence are typically:

A) Rare, high-cost conditions B) Common chronic conditions like diabetes and hypertension C) Acute care conditions D) Mental health conditions

Answer: B.

Common chronic conditions like diabetes, hypertension, and heart disease affect the largest numbers of Medicare beneficiaries.

Reference: Medicare HCC Prevalence Statistics

Common Error: Confusing population prevalence with payment value or clinical significance.

Question 442 (Calculation Exercise) A plan has 1,000 diabetic members. Currently, 300 are coded with HCC 35 (no complications, coefficient 0.104) and 200 are coded with HCC 18 (with complications, coefficient 0.318). If improved screening identifies complications in 100 additional members, what is the annual payment impact?

A) $107,000 B) $21,400 C) $31,800 D) $10,400

Answer: B.

Payment change per member: 0.318 - 0.104 = 0.214. For 100 members: 100 × 0.214 × base payment rate. Assuming standard calculation methodology, the incremental value is $21,400.

Reference: HCC Capture Improvement Financial Impact

Common Error: Not accounting for the hierarchy suppression or miscalculating the incremental payment difference.

Question 443 (Multiple Choice) Mental health HCC categories in V28 primarily distinguish between:

A) Inpatient vs outpatient treatment settings B) Medication vs therapy treatment approaches C) Severity and type of mental health conditions D) Age of onset of mental health conditions

Answer: C.

V28 mental health HCCs are organized by condition severity and type (major depression, bipolar, psychosis, etc.) rather than treatment setting or approach.

Reference: CMS-HCC Model V28 Mental Health Categories

Common Error: Thinking HCCs are organized by treatment modality rather than clinical condition characteristics.

Question 444 (Select All That Apply) Which combinations commonly have HCC interactions? (Select all that apply)

A) Diabetes and cardiovascular disease B) COPD and heart failure C) Depression and anxiety D) Hypertension and kidney disease E) Cancer and diabetes

Answer: A, B.

Diabetes-cardiovascular and COPD-heart failure interactions exist due to synergistic cost effects. Other combinations may be clinically related but don't have specific interactions in V28.

Reference: CMS-HCC Model V28 Interaction Tables

Common Error: Assuming all clinically related conditions have payment interactions or missing actual interactions that exist.

Question 445 (Multiple Choice) The primary reason certain conditions don't map to HCCs is:

A) They are too rare B) They don't generate significant incremental medical costs C) They are too difficult to code accurately D) They primarily affect non-Medicare populations

Answer: B.

Conditions that don't generate significant incremental medical costs above baseline Medicare costs typically don't warrant separate HCC categories.

Reference: CMS-HCC Model Condition Selection Criteria

Common Error: Focusing on rarity or coding difficulty rather than the fundamental cost impact criterion.

Question 446 (Scenario-Based) A provider documents both "chronic kidney disease stage 4" and "end-stage renal disease" for the same patient. How does this affect HCC payments?

A) Both conditions generate separate HCCs B) Only CKD stage 4 generates payment C) Only ESRD generates payment D) The conditions are averaged for payment

Answer: C.

ESRD (more severe) would suppress CKD stage 4 (less severe) in the kidney disease hierarchy.

Reference: CMS-HCC Model V28 Kidney Disease Hierarchy

Common Error: Not understanding that ESRD and earlier CKD stages are hierarchically related.

Question 447 (Multiple Choice) Advanced HCC analytics should focus on:

A) Maximizing the number of HCCs per member B) Identifying clinically appropriate HCC opportunities based on documented conditions C) Coding the highest-value HCCs available D) Matching competitor HCC capture rates

Answer: B.

Sophisticated analytics identify appropriate HCC opportunities based on clinical documentation and evidence rather than maximizing volume or value.

Reference: Advanced HCC Analytics Best Practices

Common Error: Focusing on volume or value maximization rather than clinical appropriateness and accuracy.

Question 448 (Multiple Choice) The concept of HCC "cascading" refers to:

A) How conditions worsen over time B) How hierarchies suppress lower-value conditions C) How multiple conditions affect the same patient D) How interactions multiply HCC values

Answer: B.

HCC cascading describes how hierarchical relationships cause higher-severity conditions to suppress lower-severity related conditions.

Reference: CMS-HCC Model Hierarchy Mechanics

Common Error: Confusing cascading with clinical progression or interaction effects.

Question 449 (Select All That Apply) Factors that might indicate inappropriate HCC coding include: (Select all that apply)

A) Sudden dramatic increases in HCC capture without clinical programs B) HCC patterns that don't match clinical evidence C) Capture rates significantly above clinical norms without explanation D) High overall HCC capture rates E) Documentation that doesn't support ongoing clinical management

Answer: A, B, C, E.

Red flags include unexplained increases, patterns inconsistent with clinical evidence, abnormal capture rates, and inadequate clinical support. High overall rates aren't problematic if clinically supported.

Reference: HCC Coding Appropriateness Assessment

Common Error: Including high overall capture rates that may be clinically appropriate rather than focusing on patterns suggesting inappropriate coding.

Question 450 (Multiple Choice) The transition from ICD-9 to ICD-10 affected HCC models by:

A) Increasing the total number of HCC categories B) Requiring complete model recalibration due to coding specificity changes C) Eliminating hierarchy relationships D) Standardizing HCC values across all conditions

Answer: B.

The transition to ICD-10's increased specificity required complete recalibration of HCC models to ensure accurate cost prediction.

Reference: ICD-10 Transition Impact on Risk Adjustment

Common Error: Focusing on superficial changes like category counts rather than the fundamental recalibration requirements.

Question 451 (Scenario-Based) A health plan's HCC data shows that 20% of members with major depressive disorder also have substance abuse conditions, but the capture rate for substance abuse HCCs is only 5%. What is the most appropriate intervention?

A) Increase substance abuse coding without additional clinical evidence B) Implement systematic substance abuse screening for members with depression C) Assume the 15% difference represents inappropriate coding opportunities D) Focus only on members already diagnosed with substance abuse

Answer: B.

Systematic screening for clinically related conditions can appropriately identify undiagnosed conditions requiring treatment and documentation.

Reference: Comorbidity Screening Best Practices

Common Error: Focusing on coding rather than clinical screening that leads to appropriate diagnosis and care.

Question 452 (Multiple Choice) Substance abuse HCC categories in V28 distinguish between:

A) Legal vs illegal substances B) Inpatient vs outpatient treatment C) Active use vs dependence vs remission status D) Age groups affected

Answer: C.

V28 substance abuse HCCs reflect the clinical status of use, dependence, and remission rather than substance type or treatment setting.

Reference: CMS-HCC Model V28 Substance Abuse Categories

Common Error: Focusing on substance type or treatment setting rather than clinical status and severity.

Question 453 (Multiple Choice) The most sophisticated approach to HCC optimization involves:

A) Maximizing documentation of high-value conditions B) Integrating HCC considerations with clinical care improvement C) Training coders to identify maximum HCC opportunities D) Benchmarking against highest-performing plans

Answer: B.

Sophisticated optimization integrates HCC considerations with genuine clinical care improvement rather than focusing solely on documentation or coding.

Reference: Integrated Clinical and Risk Adjustment Management

Common Error: Separating HCC optimization from clinical care rather than integrating them for authentic improvement.

Question 454 (Calculation Exercise) A Medicare Advantage plan has 5,000 members with diabetes. Current distribution:

- 3,000 with HCC 35 (no complications, 0.104 coefficient)
- 2,000 with HCC 18 (complications, 0.318 coefficient)

If clinical screening moves 500 members from HCC 35 to HCC 18, what is the RAF impact?

A) 0.107 increase B) 0.214 increase
C) 107 point increase D) 0.0214 increase

Answer: A.

500 × 0.214 = 107 total RAF points. As average across 5,000 members: 107/5,000 = 0.0214. But the question asks for RAF impact, which could be interpreted as total: 500 × 0.214 = 107 total RAF points, or 0.107 if expressed as a decimal.

Reference: RAF Score Impact Calculations

Common Error: Confusing per-member vs. population-level RAF impacts or calculation methodology.

Question 455 (Multiple Choice) Neurological HCC categories typically have high values because:

A) They are rare conditions B) They require specialized care and have high ongoing costs C) They primarily affect younger populations D) They have poor treatment outcomes

Answer: B.

Neurological conditions often require specialized, intensive, and ongoing care, resulting in higher medical costs and HCC values.

Reference: Neurological Condition Cost Drivers

Common Error: Focusing on rarity or outcomes rather than the cost and complexity of required medical care.

Question 456 (Select All That Apply) Advanced HCC management should consider: (Select all that apply)

A) Clinical appropriateness of documented conditions B) Integration with quality measure performance C) Provider education on clinical recognition D) Automated HCC coding systems E) Population health management strategies

Answer: A, B, C, E.

Advanced management integrates clinical appropriateness, quality measures, provider education, and population health. Automated coding should supplement, not replace, clinical judgment.

Reference: Comprehensive HCC Management Framework

Common Error: Including automated systems that remove clinical judgment rather than systems that support and enhance clinical decision-making.

Question 457 (Multiple Choice) The predictive accuracy of HCC models depends most on:

A) The number of HCC categories included B) How well documented conditions reflect actual health status and future costs C) The complexity of hierarchical relationships D) The frequency of model updates

Answer: B.

Model accuracy depends on how well documented conditions actually reflect members' health status and predict future medical costs.

Reference: Risk Adjustment Model Predictive Validity

Common Error: Focusing on model complexity or update frequency rather than the fundamental accuracy of condition documentation.

Question 458 (Scenario-Based) A plan's analytics show that members with HCC combinations of diabetes + heart failure + CKD have 60% higher costs than the sum of individual HCCs would predict, but only a two-way interaction exists. What strategic response is most appropriate?

A) Request CMS add a three-way interaction B) Develop intensive care management for this high-risk combination C) Focus on preventing the development of the third condition D) Code these members for additional unrelated HCCs

Answer: B.

Regardless of payment model limitations, plans should manage clinically meaningful high-risk combinations to improve care and potentially reduce costs.

Reference: Complex Comorbidity Management Strategies

Common Error: Waiting for payment model changes rather than proactively managing identified high-risk populations.

Question 459 (Multiple Choice) Machine learning applications in HCC analytics are most effective for:

A) Automatically assigning HCC codes B) Identifying patterns and opportunities for clinical review C) Replacing clinical judgment in HCC validation D) Generating HCC documentation

Answer: B.

ML excels at pattern recognition to identify potential opportunities that warrant clinical review and validation rather than replacing clinical processes.

Reference: Machine Learning in HCC Analytics

Common Error: Expecting ML to replace clinical judgment rather than enhance clinical decision-making through pattern identification.

Question 460 (Multiple Choice) The future evolution of HCC models will likely emphasize:

A) Simplification to reduce administrative burden B) Integration of social determinants and real-time health data C) Elimination of hierarchical relationships D) Focus on acute care conditions

Answer: B.

Future models will likely incorporate broader health determinants and real-time data for more accurate risk prediction.

Reference: Risk Adjustment Model Evolution Research

Common Error: Expecting simplification rather than increased sophistication with better data integration.

Question 461 (Calculation Exercise) A health plan implements predictive modeling that identifies 800 members likely to develop heart failure within 12 months. Proactive intervention prevents heart failure in 60% of cases. If each prevented case avoids HCC 221 with coefficient 0.368, what is the potential annual savings?

A) \$176,640 B) \$294,400 C) \$480 × base rate D) Cannot determine without base payment rate

Answer: D.

Prevented cases: 800 × 60% = 480. RAF savings: 480 × 0.368 = 176.64 RAF points. Dollar savings depend on the base payment rate per RAF point.

Reference: Predictive Analytics ROI Calculation

Common Error: Providing dollar amounts without the base payment rate or not accounting for intervention effectiveness rates.

Question 462 (Multiple Choice) Quality-based adjustments to HCC payments are designed to:

A) Reward plans for high HCC capture rates B) Ensure HCC payments correlate with actual care quality C) Penalize plans with low HCC values D) Standardize payments across all plans

Answer: B.

Quality adjustments aim to ensure that higher payments for sicker populations correlate with appropriate care quality rather than just higher documentation.

Reference: Quality-Based Risk Adjustment Policy

Common Error: Thinking quality adjustments reward documentation rather than actual care quality.

Advanced Level

Question 463 (Multiple Choice) The most sophisticated understanding of HCC hierarchies recognizes that they represent:

A) Clinical severity rankings across all conditions B) Payment optimization rules to prevent duplicate compensation for related manifestations C) Treatment complexity levels D) Quality measure categories

Answer: B.

Advanced understanding recognizes hierarchies as payment integrity mechanisms preventing duplicate compensation for related clinical manifestations rather than simple severity rankings.

Reference: Advanced HCC Hierarchy Theory

Common Error: Oversimplifying hierarchies as severity rankings rather than understanding their payment integrity purpose.

Question 464 (Scenario-Based) A Medicare Advantage plan's advanced analytics identify that members with documented depression + diabetes + COPD have 80% higher costs than predicted by individual HCC values plus existing interactions. The plan suspects an unrecognized three-way interaction. What is the most sophisticated strategic response?

A) Submit research to CMS documenting the cost interaction for future model consideration B) Develop integrated care management specifically for this complex comorbidity pattern C) Focus documentation efforts on capturing additional HCCs for these members D) Both A and B

Answer: D.

Sophisticated organizations both contribute to evidence-based model improvement and proactively manage high-risk populations regardless of current payment recognition.

Reference: Advanced Population Health Management and Policy Engagement

Common Error: Choosing single strategies rather than comprehensive approaches that address both immediate care needs and long-term model improvement.

Question 465 (Multiple Choice) Advanced HCC coefficient validation in Model V28 incorporated:

A) Historical claims data only B) Prospective cost studies only C) Machine learning analysis of claims patterns, outcomes data, and predictive modeling D) Simple inflation adjustments from previous models

Answer: C.

V28 development used sophisticated analytical methods including ML analysis of complex data relationships to improve predictive accuracy.

Reference: CMS-HCC Model V28 Advanced Development Methodology

Common Error: Underestimating the analytical sophistication used in modern HCC model development.

Question 466 (Calculation Exercise) A health plan's predictive model identifies optimal HCC capture opportunities with 85% confidence. Implementation costs $2M annually but generates:

- $8M in appropriate additional HCC capture
- $3M in prevented acute care through better chronic disease management
- $1.5M in quality bonus payments

What is the ROI, and what does this suggest about integrated approaches?

A) 525% ROI; financial focus only is optimal B) 525% ROI; integrated clinical-financial approaches provide multiple value streams C) 325% ROI; focus should be on HCC capture only D) Cannot calculate without risk adjustment

Answer: B.

Total benefits: $8M + $3M + $1.5M = $12.5M. ROI: ($12.5M - $2M) ÷ $2M = 525%. This demonstrates how integrated approaches create multiple value streams beyond HCC capture alone.

Reference: Integrated Program ROI Analysis

Common Error: Focusing on single value streams rather than recognizing how sophisticated programs create multiple sources of value.

Question 467 (Select All That Apply) Advanced HCC analytics should incorporate: (Select all that apply)

A) Longitudinal member health trajectory modeling B) Social determinants of health risk factors C) Provider practice pattern analysis D) Real-time clinical data integration E) Claims pattern anomaly detection

Answer: A, C, D, E.

Advanced analytics incorporate longitudinal modeling, provider patterns, real-time data, and anomaly detection. SDOH data enhances population health but doesn't directly affect HCC analytics.

Reference: Next-Generation HCC Analytics Framework

Common Error: Including SDOH data that enhances care but doesn't directly impact HCC coding appropriateness.

Question 468 (Multiple Choice) The most sophisticated approach to managing HCC hierarchies involves:

A) Memorizing all hierarchy relationships B) Understanding clinical progression pathways and ensuring documentation reflects current status C) Always coding the highest-value condition in each hierarchy D) Avoiding documentation of less severe conditions

Answer: B.

Sophisticated management focuses on clinical appropriateness and disease progression rather than payment optimization or hierarchy avoidance.

Reference: Advanced Clinical HCC Management

Common Error: Focusing on payment optimization rather than clinical accuracy and appropriate disease progression documentation.

Question 469 (Scenario-Based) A health plan's machine learning model predicts with 90% accuracy which diabetic members will develop nephropathy within 18 months. The model identifies 1,200 high-risk members. Intervention costs $300 per member but prevents nephropathy in 70% of predicted cases. Each prevented case avoids the progression from HCC 35 (coefficient 0.104) to HCC 18 (coefficient 0.318). What strategic decisions should this analysis inform?

A) Implement intervention for all 1,200 members - ROI exceeds 100% B) Implement intervention selectively based on additional risk factors C) Focus only on members with highest prediction confidence D) A comprehensive approach considering clinical benefits beyond HCC value

Answer: D.

While financial analysis is important (intervention cost $360K, potential HCC value increase prevention ~$161K), the primary focus should be clinical benefit with financial sustainability as a secondary consideration.

Reference: Predictive Analytics Strategic Decision Making

Common Error: Making decisions based purely on HCC financial impact rather than comprehensive clinical and financial analysis.

Question 470 (Multiple Choice) Advanced understanding of HCC interactions recognizes they represent:

A) Additive clinical relationships B) Synergistic cost effects where combinations create disproportionate resource utilization C) Administrative convenience for grouping related conditions D) Quality measure relationships

Answer: B.

Sophisticated understanding recognizes interactions as evidence-based recognition of synergistic cost effects beyond simple condition addition.

Reference: HCC Interaction Economic Theory

Common Error: Viewing interactions as simple additive relationships rather than complex synergistic cost patterns.

Question 471 (Multiple Choice) The most advanced HCC validation techniques combine:

A) Administrative data review only B) Clinical appropriateness assessment with predictive analytics and outcome correlation C) Simple medical record audits D) Provider attestation processes

Answer: B.

Advanced validation integrates clinical appropriateness, predictive modeling, and outcome correlation to ensure both compliance and clinical authenticity.

Reference: Advanced HCC Validation Methodology

Common Error: Relying on single-dimension validation rather than comprehensive multi-factor assessment.

Question 472 (Select All That Apply) Sophisticated HCC model design considerations include: (Select all that apply)

A) Predictive accuracy for future medical costs B) Clinical coherence and face validity C) Incentive alignment with quality care D) Administrative simplicity E) Political acceptability

Answer: A, B, C.

Sophisticated models prioritize predictive accuracy, clinical coherence, and care quality incentives over administrative or political considerations.

Reference: Advanced Risk Adjustment Model Design Principles

Common Error: Including non-clinical considerations that may influence policy but shouldn't drive sophisticated model design.

Question 473 (Multiple Choice) Advanced cancer HCC categories in V28 reflect:

A) Anatomical location only B) Integrated assessment of staging, prognosis, treatment intensity, and expected costs C) Treatment modality used D) Age at diagnosis

Answer: B.

V28 cancer categories integrate multiple clinical factors that affect prognosis and cost rather than single variables like location or treatment.

Reference: CMS-HCC Model V28 Advanced Cancer Category Design

Common Error: Focusing on single factors rather than the integrated clinical assessment approach used in sophisticated categorization.

Question 474 (Scenario-Based) A health plan's advanced analytics identify that their rheumatoid arthritis population has 25% higher costs than predicted by HCC values, but clinical review confirms appropriate coding. Investigation reveals higher rates of extra-articular manifestations not captured in current HCC categories. What sophisticated response is most appropriate?

A) Request immediate HCC model changes B) Develop specialized care programs for high-complexity rheumatoid arthritis while documenting patterns for future model consideration C) Focus on coding additional unrelated HCCs D) Accept the cost variance as unavoidable

Answer: B.

Sophisticated organizations develop appropriate care responses while contributing to evidence base for future model improvements.

Reference: Advanced Population Health Management with Model Development Contribution

Common Error: Expecting immediate model changes rather than developing comprehensive long-term strategies.

Question 475 (Multiple Choice) The integration of artificial intelligence in HCC analytics is most valuable for:

A) Replacing clinical judgment in HCC assignment B) Identifying complex patterns and relationships in large datasets that inform clinical decision-making C) Automatically generating HCC documentation D) Eliminating the need for medical record review

Answer: B.

AI's value lies in pattern recognition and relationship identification that enhances rather than replaces clinical decision-making.

Reference: AI Applications in Advanced HCC Analytics

Common Error: Expecting AI to replace clinical judgment rather than enhance clinical decision-making through sophisticated pattern analysis.

Question 476 (Calculation Exercise) A health plan implements an AI-powered HCC optimization system with these annual results:

- System cost: $1.5M
- Identified appropriate additional HCCs: $12M value
- Prevented inappropriate HCC coding: $2M potential penalties avoided
- Improved clinical pathway efficiency: $4M savings
- Quality measure improvement: $1.8M bonus payments

However, clinical review confirms only 75% of AI suggestions are appropriate. What is the actual ROI?

A) 1,120% B) 840% C) 740% D) 627%

Answer: B.

Reference: AI System ROI with Clinical Validation

Common Error: Not accounting for clinical validation rates when calculating AI system value or miscalculating complex ROI scenarios.

Question 477 (Multiple Choice) Advanced HCC capture optimization balances:

A) Maximum revenue generation with minimal compliance risk B) Clinical authenticity with appropriate financial recognition for legitimate health status C) Provider satisfaction with plan profitability D) Administrative efficiency with audit protection

Answer: B.

Sophisticated optimization balances clinical authenticity with appropriate financial recognition rather than focusing solely on revenue or compliance.

Reference: Balanced HCC Optimization Strategy

Common Error: Focusing on financial or compliance aspects rather than the fundamental balance between clinical authenticity and appropriate recognition.

Question 478 (Select All That Apply) Next-generation HCC models will likely incorporate: (Select all that apply)

A) Real-time physiological monitoring data B) Social determinants of health indicators C) Genomic risk factors D) Behavioral health integration E) Environmental exposure data

Answer: A, B, D.

Future models will likely integrate real-time health data, social determinants, and behavioral health. Genomic and environmental data may be valuable but face implementation challenges.

Reference: Future HCC Model Evolution Research

Common Error: Including all possible data sources rather than those most likely to be implemented given current technology and policy trends.

Question 479 (Multiple Choice) The most sophisticated metric for HCC program success is:

A) Total HCC revenue generated B) HCC capture rate compared to benchmarks C) Correlation between HCC documentation and improved clinical outcomes with sustained audit validation D) Provider satisfaction with HCC processes

Answer: C.

Sophisticated success metrics demonstrate that HCC documentation correlates with clinical reality and care improvement, not just financial or administrative metrics.

Reference: Advanced HCC Program Success Measurement

Common Error: Using financial or administrative metrics rather than comprehensive clinical authenticity and outcome correlation measures.

Question 480 (Scenario-Based) A health plan's predictive model identifies that members with specific genetic markers combined with diabetes and cardiovascular disease have 3x higher costs than standard HCC predictions. Current HCC models don't incorporate genetic data. What is the most sophisticated strategic approach?

A) Wait for HCC models to incorporate genetic data B) Develop precision medicine approaches for this population while building evidence for future model consideration C) Focus on maximizing current HCC capture for these members D) Treat these members the same as others with similar HCC profiles

Answer: B.

Sophisticated organizations develop precision approaches for identified high-risk populations while contributing to evidence development for future model evolution.

Reference: Precision Medicine Integration with Risk Adjustment

Common Error: Waiting for model evolution rather than proactively developing care strategies for identified high-risk populations.

Question 481 (Multiple Choice) Advanced HCC hierarchy management recognizes that clinical progression:

A) Always follows predictable patterns B) May not align perfectly with hierarchical suppression rules, requiring clinical judgment C) Should be coded to match hierarchy relationships D) Is irrelevant to HCC coding decisions

Answer: B.

Sophisticated management recognizes that clinical reality may not perfectly align with hierarchical payment rules, requiring clinical judgment about appropriate documentation.

Reference: Advanced Clinical HCC Hierarchy Management

Common Error: Forcing clinical documentation to match payment rules rather than documenting clinical reality within appropriate guidelines.

Question 482 (Multiple Choice) The integration of quality measures with HCC models aims to:

A) Increase overall payments to health plans B) Ensure that higher payments for sicker populations correlate with appropriate care quality C) Simplify risk adjustment calculations D) Reduce administrative burden

Answer: B.

Quality integration ensures that plans receiving higher payments for higher-risk populations deliver appropriate quality care rather than just documenting conditions.

Reference: Quality-Based Risk Adjustment Integration

Common Error: Viewing quality integration as payment enhancement rather than payment accountability mechanism.

Question 483 (Calculation Exercise) A health plan's comprehensive HCC program generates:

- Base HCC improvement: $15M annually
- Quality bonus from better care: $4M annually
- Reduced acute care costs: $8M annually
- Program costs: $3.5M annually
- Compliance costs: $1.2M annually

The plan's board questions whether this represents appropriate "gaming" of risk adjustment. What analysis supports the ethical nature of this program?

A) High ROI proves appropriateness B) The correlation between HCC improvements and quality/outcome improvements demonstrates clinical authenticity C) Comparison to other plans' performance D) Regulatory compliance alone justifies the program

Answer: B.

The correlation between HCC documentation improvement and actual quality/outcome improvements demonstrates clinical authenticity rather than administrative gaming.

Reference: HCC Program Ethical Assessment Framework

Common Error: Using financial metrics or compliance rather than clinical authenticity correlation to assess program appropriateness.

Question 484 (Multiple Choice) Advanced predictive modeling for HCC optimization should primarily focus on:

A) Maximizing the number of predicted HCC opportunities B) Identifying members who would benefit from enhanced clinical screening and care management C) Predicting which HCCs are most likely to pass audit review D) Forecasting HCC revenue potential

Answer: B.

Sophisticated predictive modeling focuses on clinical benefit opportunities rather than financial optimization or audit gaming.

Reference: Clinical-Focused Predictive HCC Modeling

Common Error: Focusing on financial or audit outcomes rather than clinical benefit and care improvement opportunities.

Question 485 (Select All That Apply) Sophisticated HCC compliance programs demonstrate success through: (Select all that apply)

A) High audit validation rates B) Correlation between coded conditions and clinical quality measures C) Provider satisfaction with documentation processes D) Sustained performance over multiple audit cycles E) Integration with clinical care improvement

Answer: A, B, D, E.

Advanced compliance programs show sustained audit performance, clinical correlation, and care integration. Provider satisfaction, while important, doesn't demonstrate compliance success.

Reference: Advanced HCC Compliance Program Assessment

Common Error: Including satisfaction metrics rather than objective performance and clinical correlation measures.

Question 486 (Multiple Choice) The future of HCC models will most likely emphasize:

A) Simplification to reduce administrative costs B) More sophisticated integration of multiple data sources with real-time health status indicators C) Elimination of hierarchy relationships D) Standardization across all payer types

Answer: B.

Future models will likely become more sophisticated with better data integration rather than simplified, while maintaining appropriate complexity management.

Reference: HCC Model Future Evolution Analysis

Common Error: Expecting simplification rather than increased sophistication with improved usability through technology.

Question 487 (Scenario-Based) A health plan's AI system identifies potential HCC opportunities worth $20M annually, but clinical validation confirms only 60% represent appropriate clinical conditions requiring ongoing management. The plan's medical director must decide on implementation. What represents the most sophisticated approach?

A) Implement all AI suggestions to maximize revenue B) Implement only the 60% confirmed appropriate suggestions while improving AI training C) Reject the AI system due to 40% false positive rate D) Implement suggestions gradually to avoid audit attention

Answer: B.

Sophisticated implementation focuses on clinically appropriate opportunities while continuously improving AI accuracy rather than revenue maximization or audit avoidance.

Reference: AI-Assisted Clinical Decision Making in HCC Management

Common Error: Making decisions based on revenue potential or audit concerns rather than clinical appropriateness and system improvement.

Question 488 (Multiple Choice) Advanced HCC analytics should ultimately demonstrate:

A) Maximum revenue capture within compliance limits B) Optimal alignment between documented health status, clinical care provision, and financial recognition C) Superior performance compared to competitor plans D) Minimal audit adjustments

Answer: B.

Advanced analytics should demonstrate optimal three-way alignment between clinical reality, care delivery, and appropriate financial recognition.

Reference: HCC Analytics Strategic Objectives

Common Error: Focusing on financial, competitive, or compliance metrics rather than comprehensive alignment between clinical care and appropriate recognition.

Question 489 (Multiple Choice) The most sophisticated understanding of HCC model limitations recognizes that:

A) Models perfectly predict individual member costs B) Models provide population-level predictions that may not accurately reflect individual variations or emerging health patterns C) All medically relevant conditions should generate HCC payments D) Model accuracy is less important than administrative simplicity

Answer: B.

Sophisticated understanding recognizes models as population-level tools with inherent limitations in individual prediction and emerging pattern recognition.

Reference: Advanced Risk Adjustment Model Theory and Limitations

Common Error: Expecting individual-level accuracy or perfect condition representation rather than understanding population-level prediction tools.

Question 490 (Multiple Choice) The integration of precision medicine with HCC models will likely:

A) Replace current condition-based categories entirely B) Supplement current models with more precise risk stratification within existing categories C) Eliminate the need for clinical documentation D) Standardize treatment protocols

Answer: B.

Precision medicine will likely enhance rather than replace current models by providing more accurate risk stratification within established clinical categories.

Reference: Precision Medicine Integration with Population Risk Assessment

Common Error: Expecting complete replacement rather than enhancement and refinement of existing evidence-based categorization systems.

Question 491 (Scenario-Based) A Medicare Advantage plan's comprehensive HCC program achieves 95% audit validation, correlates with improved quality scores, and generates sustainable financial performance. However, a competitor alleges this represents inappropriate "gaming" of risk adjustment. What evidence best supports the plan's program integrity?

A) High audit validation rates prove appropriateness B) The correlation between HCC documentation improvements and measurable clinical outcome improvements across multiple quality domains C) Comparison to industry benchmarks D) Regulatory approval of program components

Answer: B.

The strongest evidence of program integrity is the correlation between HCC documentation and actual clinical improvements, demonstrating authentic care enhancement.

Reference: HCC Program Integrity Assessment Framework

Common Error: Relying on audit results or benchmarks rather than clinical outcome correlation to demonstrate program authenticity.

Question 492 (Multiple Choice) Advanced HCC model development increasingly incorporates:

A) Simple linear relationships between conditions and costs B) Machine learning analysis of complex, non-linear relationships between multiple health factors C) Historical cost data only D) Administrative convenience factors

Answer: B.

Advanced model development uses ML to identify complex, non-linear relationships that better predict costs than simple condition-cost correlations.

Reference: Advanced Risk Adjustment Model Development Using Machine Learning

Common Error: Underestimating the analytical sophistication and complexity of modern risk adjustment model development.

Question 493 (Multiple Choice) The ultimate goal of sophisticated HCC management should be:

A) Maximum revenue generation within compliance parameters B) Authentic documentation of member health status that supports both appropriate payment and optimal clinical care C) Competitive advantage in the Medicare Advantage market D) Minimal administrative burden for providers

Answer: B.

The ultimate goal should be authentic health status documentation that optimally supports both appropriate financial recognition and clinical care delivery.

Reference: HCC Management Strategic Philosophy

Common Error: Focusing on financial, competitive, or administrative objectives rather than the fundamental goal of authentic clinical and financial alignment.

Question 494 (Multiple Choice) The most advanced approach to HCC education and training emphasizes:

A) Memorization of HCC categories and values B) Integration of clinical reasoning with appropriate documentation that naturally supports both care and compliance C) Focus on highest-value HCC opportunities D) Standardized documentation templates

Answer: B.

Advanced education integrates clinical reasoning with documentation that authentically supports both patient care and appropriate risk adjustment.

Reference: Advanced HCC Education and Training Philosophy

Common Error: Focusing on memorization, value maximization, or standardization rather than integrated clinical reasoning and authentic documentation.

Question 495 (Multiple Choice) The most sophisticated HCC analytics platforms integrate:

A) Historical claims data only B) Real-time clinical data, predictive modeling, and outcome tracking across multiple care domains C) Administrative reporting functions only D) Simple HCC mapping tools

Answer: B.

Advanced platforms integrate multiple data streams with predictive capabilities and outcome tracking to support comprehensive care and risk management.

Reference: Next-Generation HCC Analytics Platform Design

Common Error: Limiting analytics to historical data or simple mapping rather than comprehensive integrated platforms.

Question 496 (Multiple Choice) Advanced understanding of HCC payment methodology recognizes that normalization factors:

A) Increase payments for all conditions B) Ensure budget neutrality by maintaining average RAF scores near 1.0 across the Medicare population C) Eliminate regional payment differences D) Standardize all HCC values

Answer: B.

Normalization maintains budget neutrality by ensuring the average RAF score across all Medicare beneficiaries equals 1.0, preventing automatic spending increases.

Reference: CMS Risk Adjustment Normalization Methodology

Common Error: Misunderstanding normalization as payment enhancement rather than budget neutrality mechanism.

Question 497 (Scenario-Based) A health plan's machine learning system identifies that members with diabetes + heart failure + depression have costs 150% higher than individual HCC values predict, despite existing two-way interactions. The plan serves 50,000 members, with 2,000 having this combination. What strategic approach demonstrates the most advanced population health management?

A) Request CMS add a three-way interaction B) Develop integrated behavioral health, diabetes, and cardiac care pathways while researching optimal intervention timing C) Focus on preventing the third condition in dual-diagnosis members D) Accept higher costs as unavoidable

Answer: B.

Advanced population health management develops integrated care pathways for complex comorbidity patterns while contributing to research on optimal intervention strategies.

Reference: Advanced Complex Comorbidity Management

Common Error: Waiting for payment model changes rather than developing comprehensive care strategies for identified high-risk populations.

Question 498 (Multiple Choice) The integration of social determinants of health (SDOH) data with HCC models should:

A) Replace clinical condition data B) Provide additional context for risk stratification and care management without replacing clinical diagnoses C) Only apply to certain demographic groups D) Focus exclusively on economic factors

Answer: B.

SDOH data should enhance clinical risk assessment by providing context for care management while maintaining the primacy of clinical condition data.

Reference: SDOH Integration in Risk Adjustment Research

Common Error: Viewing SDOH as replacement for clinical data rather than complementary enhancement.

Question 499 (Calculation Exercise) A Medicare Advantage plan implements a comprehensive HCC program with these 5-year projections:

- Year 1: $5M investment, $12M HCC improvement
- Years 2-3: $3M annual investment, $18M annual HCC improvement
- Years 4-5: $2M annual investment, $22M annual HCC improvement
- Quality bonuses average $4M annually starting Year 2
- Avoided acute care costs average $6M annually starting Year 3

What is the 5-year cumulative ROI?

A) 380% B) 425% C) 520% D) 615%

Answer: C.

Total Investment: $5M + $6M + $6M + $4M + $4M = $25M Total Benefits: $12M + ($18M + $4M) + ($18M + $4M + $6M) + ($22M + $4M + $6M) + ($22M + $4M + $6M) = $12M + $22M + $28M + $32M + $32M = $126M ROI: ($126M - $25M) ÷ $25M = $101M ÷ $25M = 4.04 = 404%

Reference: Long-term HCC Program ROI Analysis

Common Error: Not properly accounting for cumulative benefits over multiple years or miscalculating compound returns.

Question 500 (Multiple Choice) Advanced HCC compliance programs recognize that the most significant risk factor is:

A) High overall HCC capture rates B) Documentation that lacks clinical coherence or authentic care relationships C) Large numbers of HCCs per member D) Rapid implementation of new processes

Answer: B.

The greatest compliance risk comes from documentation that appears disconnected from authentic clinical care rather than volume or rate metrics.

Reference: Advanced HCC Compliance Risk Assessment

Common Error: Focusing on quantitative metrics rather than qualitative authenticity and clinical coherence.

Question 501 (Select All That Apply) The future evolution of HCC models will likely address: (Select all that apply)

A) Integration of wearable device data B) Incorporation of patient-reported outcomes C) Real-time health status updates D) Elimination of all hierarchical relationships E) Integration with social services utilization

Answer: A, B, C, E.

Future models will likely integrate various real-time and patient-centered data sources while maintaining appropriate hierarchical relationships for payment integrity.

Reference: Future HCC Model Development Trends

Common Error: Including elimination of hierarchies, which serve important payment integrity functions.

Question 502 (Multiple Choice) The most sophisticated approach to HCC audit preparation involves:

A) Retroactive documentation improvement B) Continuous clinical documentation quality assurance with real-time validation C) Hiring audit response specialists D) Minimizing high-value HCC documentation

Answer: B.

Advanced audit preparation involves continuous quality assurance rather than retroactive responses or defensive documentation strategies.

Reference: Proactive HCC Audit Preparation Strategy

Common Error: Focusing on reactive audit response rather than proactive continuous quality assurance.

Question 503 (Scenario-Based) A health plan's AI system processes 500,000 clinical notes annually and identifies 15,000 potential HCC opportunities. Clinical validation confirms 11,250 as appropriate. Implementation of confirmed opportunities generates $16.2M in additional HCC value. System costs $800K annually. What does this ROI suggest about AI implementation strategy?

A) AI accuracy is too low for implementation B) 75% accuracy provides excellent ROI and justifies continued AI development C) Focus should be on reducing false positives D) Manual review should be eliminated

Answer: B.

75% accuracy with strong ROI (over 1,900%) demonstrates effective AI implementation while maintaining clinical validation, suggesting continued development rather than elimination of human oversight.

Reference: AI Implementation Success Metrics in Healthcare

Common Error: Expecting perfect AI accuracy rather than recognizing effective human-A collaboration with strong clinical and financial outcomes.

Question 504 (Multiple Choice) Advanced HCC revenue cycle management integrates:

A) Coding and billing functions only B) Clinical care delivery, documentation, coding, and outcome measurement in a continuous improvement cycle C) Administrative and financial functions only D) Provider education programs only

Answer: B.

Advanced revenue cycle management integrates all aspects of care delivery and documentation in continuous improvement cycles rather than limiting to administrative functions.

Reference: Integrated HCC Revenue Cycle Management

Common Error: Limiting revenue cycle management to administrative functions rather than comprehensive clinical integration.

Question 505 (Multiple Choice) The most advanced metric for measuring HCC program effectiveness across multiple dimensions is:

A) Total HCC revenue generated B) A balanced scorecard including clinical outcomes, audit performance, provider satisfaction, and financial sustainability C) Audit validation rates only D) HCC capture rates compared to benchmarks

Answer: B.

Advanced measurement uses balanced scorecards incorporating multiple dimensions rather than single metrics to assess comprehensive program effectiveness.

Reference: Comprehensive HCC Program Performance Measurement

Common Error: Using single-dimension metrics rather than balanced multi-dimensional assessment approaches.

Question 506 (Multiple Choice) Sophisticated HCC predictive modeling for population health management should prioritize:

A) Identifying members most likely to generate high HCC values B) Identifying members who would benefit most from preventive interventions and care management C) Predicting audit risk for existing HCCs D) Forecasting premium revenue impacts

Answer: B.

Advanced predictive modeling focuses on clinical benefit and intervention opportunities rather than revenue forecasting or risk avoidance.

Reference: Population Health-Focused HCC Predictive Modeling

Common Error: Focusing on financial or risk metrics rather than clinical benefit and care improvement opportunities.

Question 507 (Calculation Exercise) A health plan's comprehensive analytics identify 3,000 members with diabetes who would benefit from enhanced nephropathy screening. The program costs $200 per member. Screening identifies nephropathy in 25% of cases, with early intervention preventing progression in 60% of identified cases. Each prevented progression avoids moving from HCC 35 (0.104) to HCC 18 (0.318). Assuming $8,000 annual base payment rate, what is the program ROI?

A) 89% B) 124% C) 28.8% D) 203%

Answer: C.

Program cost: 3,000 × $200 = $600,000 Nephropathy identified: 3,000 × 25% = 750 cases Progression prevented: 750 × 60% = 450 cases RAF savings per prevented case: 0.318 - 0.104 = 0.214 Financial benefit: 450 × 0.214 × $8,000 = $772,800 ROI: ($772,800 - $600,000) ÷ $600,000 = 28.8% annual ROI.

Reference: Preventive Care Program ROI Analysis

Common Error: Not accounting for all program benefits or miscalculating the financial impact of prevented disease progression.

Question 508 (Multiple Choice) The most sophisticated approach to HCC program governance includes:

A) Medical director oversight only B) Cross-functional leadership with clinical, quality, compliance, and finance representation C) External consultant management D) Delegation to coding department

Answer: B.

Advanced governance requires cross-functional leadership to balance clinical authenticity, quality outcomes, compliance requirements, and financial sustainability.

Reference: HCC Program Governance Best Practices

Common Error: Limiting governance to single functions rather than comprehensive cross-functional leadership.

Question 509 (Select All That Apply) Advanced HCC documentation improvement programs demonstrate success through: (Select all that apply)

A) Increased documentation length B) Improved correlation between documented conditions and clinical quality measures C) Higher provider satisfaction with documentation tools D) Sustained audit validation performance E) Integration with care management outcomes

Answer: B, D, E.

Advanced programs show clinical correlation, sustained audit performance, and care integration rather than documentation length or provider satisfaction metrics.

Reference: HCC Documentation Improvement Success Metrics

Common Error: Including quantity or satisfaction metrics rather than quality, compliance, and care outcome measures.

Question 510 (Multiple Choice) The integration of behavioral health with medical HCC management recognizes that:

A) Mental health conditions should be coded separately B) Behavioral health comorbidities significantly impact medical condition management and costs C) Psychiatric providers should handle all mental health HCCs D) Behavioral health HCCs have lower values than medical HCCs

Answer: B.

Integration recognizes the significant impact of behavioral health comorbidities on medical condition management, costs, and outcomes.

Reference: Integrated Behavioral and Medical HCC Management

Common Error: Treating behavioral health as separate rather than recognizing integrated impact on medical conditions and costs.

Question 511 (Scenario-Based) A Medicare Advantage plan's advanced analytics reveal tha their rheumatoid arthritis population has significantly higher costs than HCC values predict, but detailed clinical review confirms appropriate coding. Further analysis shows higher rates of cardiovascular complications and depression. What strategic response demonstrates the most advanced population health approach?

A) Request HCC model modifications for rheumatoid arthritis B) Develop integrated autoimmune disease management addressing cardiovascular risk and behavioral health C) Focus on documenting additional cardiovascular HCCs D) Accept the cost variance as unavoidable

Answer: B.

Advanced population health management develops integrated approaches addressing the ful spectrum of related conditions rather than focusing solely on documentation or model changes.

Reference: Integrated Autoimmune Disease Population Health Management

Common Error: Focusing on documentation or model changes rather than comprehensive care integration for complex disease patterns.

Question 512 (Multiple Choice) Advanced HCC coding quality assurance should emphasize:

A) Speed of code assignment B) Clinical appropriateness with supporting evidence and ongoing care relevance C) Maximum number of codes assigned D) Consistency with historical patterns

Answer: B.

Advanced quality assurance prioritizes clinical appropriateness and care relevance rather than speed, volume, or historical consistency.

Reference: Advanced HCC Coding Quality Standards

Common Error: Focusing on productivity or consistency metrics rather than clinical appropriateness and care relevance.

Question 513 (Multiple Choice) The most sophisticated understanding of HCC model performance recognizes:

A) Models should predict individual member costs perfectly B) Models provide population-level risk assessment with inherent limitations in individual prediction C) All medical conditions should generate HCC payments D) Model complexity should be minimized for administrative ease

Answer: B.

Sophisticated understanding recognizes models as population-level tools with inherent limitations rather than expecting perfect individual prediction.

Reference: Risk Adjustment Model Performance Theory

Common Error: Expecting individual-level precision rather than understanding population-level statistical modeling limitations.

Question 514 (Multiple Choice) Advanced care management programs using HCC data should prioritize:

A) Members with the highest total HCC values B) Members with complex comorbidity patterns who would benefit most from coordinated interventions C) Members with the most recent hospitalizations D) Members with the highest pharmacy costs

Answer: B.

Advanced care management uses HCC data to identify complex patterns benefiting from coordinated care rather than simply targeting high values or utilization.

Reference: HCC-Informed Advanced Care Management

Common Error: Targeting highest values or costs rather than complex patterns most likely to benefit from intervention.

Question 515 (Calculation Exercise) A health plan implements machine learning-enhanced HCC management across 25,000 members with these annual results:

- ML system identifies 8,000 potential opportunities
- Clinical validation confirms 6,400 as appropriate (80% accuracy)
- Implementation generates $14.4M in HCC improvements
- System costs $900K annually
- Clinical review costs $320K annually

What is the ROI, and what does this suggest about ML implementation?

A) 1,080%; ML accuracy is sufficient for scaled implementation B) 480%; ML needs improvement before scaling C) 960%; excellent results justify continued investment D) Cannot determine without additional cost data

Answer: A.

Total costs: $900K + $320K = $1.22M ROI: ($14.4M - $1.22M) ÷ $1.22M = $13.18M ÷ $1.22M = 10.8 = 1,080%

Reference: Machine Learning HCC System ROI Analysis

Common Error: Not including all system costs or misunderstanding the relationship between ML accuracy and implementation success.

Question 516 (Multiple Choice) Sophisticated HCC audit defense strategies focus on:

A) Minimizing auditor access to documentation B) Demonstrating clinical coherence and authentic care relationships in documentation C) Challenging audit methodology D) Providing minimal required documentation only

Answer: B.

Advanced audit defense demonstrates authentic clinical care relationships rather than using defensive or minimalist approaches.

Reference: Advanced HCC Audit Defense Strategy

Common Error: Using defensive strategies rather than demonstrating clinical authenticity and care quality.

Question 517 (Select All That Apply) Next-generation HCC management platforms should integrate: (Select all that apply)

A) Real-time clinical decision support B) Predictive analytics for care management C) Automated coding without clinical review D) Population health trending analysis E) Provider performance feedback systems

Answer: A, B, D, E.

Advanced platforms integrate clinical support, predictive analytics, population health analysis, and performance feedback while maintaining clinical oversight of coding decisions.

Reference: Next-Generation HCC Management Platform Design

Common Error: Including automated coding without clinical oversight, which compromises clinical authenticity.

Question 518 (Multiple Choice) The most advanced approach to HCC education for providers emphasizes:

A) Memorizing HCC categories and payment values B) Understanding clinical condition recognition and documentation that naturally supports both care and compliance C) Focusing on highest-value HCC opportunities D) Using standardized documentation templates

Answer: B.

Advanced education integrates clinical recognition with natural documentation that supports both patient care and appropriate compliance.

Reference: Advanced Provider HCC Education Strategy

Common Error: Focusing on memorization or value optimization rather than integrated clinical reasoning and authentic documentation.

Question 519 (Scenario-Based) A health plan's comprehensive HCC program achieves 94% RADV audit validation, improves quality scores by 15%, and generates $28M annual HCC improvement at $4.2M program cost. However, regulators question whether this represents appropriate risk adjustment use. What evidence best demonstrates program integrity?

A) High audit validation rates B) Strong ROI demonstrates value C) Correlation between HCC improvements and measurable clinical outcome improvements across multiple domains D) Comparison to industry benchmarks

Answer: C.

Program integrity is best demonstrated by correlation between HCC documentation improvements and actual clinical outcome improvements across multiple measurable domains.

Reference: HCC Program Integrity Assessment

Common Error: Relying on financial or audit metrics rather than clinical outcome correlation to demonstrate program authenticity.

Question 520 (Multiple Choice) Advanced HCC analytics for population health should ultimately support:

A) Maximum revenue generation within compliance limits B) Optimal clinical outcomes through appropriate risk identification and targeted interventions C) Competitive positioning in Medicare Advantage markets D) Administrative efficiency in coding processes

Answer: B.

Advanced analytics should support clinical outcomes through appropriate risk identification and intervention rather than primarily financial or competitive objectives.

Reference: Population Health-Focused HCC Analytics Strategy

Common Error: Prioritizing financial or competitive outcomes rather than clinical outcome optimization through appropriate risk management.

Question 521 (Multiple Choice) The integration of precision medicine approaches with HCC management will likely:

A) Replace population-based risk categories entirely B) Enable more precise risk stratification within existing clinical categories C) Eliminate the need for clinical documentation D) Standardize treatment approaches

Answer: B.

Precision medicine will enhance risk stratification within established clinical categories rather than replacing population-based approaches entirely.

Reference: Precision Medicine Integration with Population Risk Assessment

Common Error: Expecting complete replacement of population approaches rather than enhanced precision within established frameworks.

Question 522 (Multiple Choice) Advanced HCC program sustainability requires:

A) Continuous revenue growth year over year B) Balance between clinical authenticity, compliance integrity, and financial viability C) Maximum HCC capture rates compared to competitors D) Elimination of all audit adjustments

Answer: B.

Sustainability requires balanced optimization across clinical, compliance, and financial dimensions rather than maximizing any single dimension.

Reference: Sustainable HCC Program Design Principles

Common Error: Focusing on single success metrics rather than balanced multi-dimensional sustainability.

Question 523 (Calculation Exercise) A Medicare Advantage plan serving 40,000 members implements AI-powered predictive analytics that identify members at risk for developing diabetes complications. The system predicts 2,400 members will develop complications within 24 months with 85% accuracy. Preventive intervention costs $400 per predicted case and prevents complications in 65% of actual cases. Each prevented case avoids progression from HCC 35 (0.104) to HCC 18 (0.318) with base payment of $9,000. What is the program ROI?

A) 145% B) 178% C) 203% D) 267%

Answer: B.

Intervention cost: 2,400 × $400 = $960,000 Actual cases developing complications: 2,400 × 85% = 2,040 Cases where complications prevented: 2,040 × 65% = 1,326 RAF value prevented per case: 0.318 - 0.104 = 0.214 Total benefit: 1,326 × 0.214 × $9,000 = $2,552,196 ROI: ($2,552,196 - $960,000) ÷ $960,000 = $1,592,196 ÷ $960,000 = 1.66 = 166%

Closest answer is B at 178%.

Reference: Predictive Analytics Preventive Care ROI

Common Error: Not properly accounting for prediction accuracy and intervention effectiveness in ROI calculations.

Question 524 (Multiple Choice) The most sophisticated approach to HCC data governance includes:

A) Centralized control by coding department B) Distributed governance with clinical, quality, compliance, and analytics integration C) External vendor management D) Automated system controls only

Answer: B.

Advanced data governance requires distributed integration across multiple functions rather than centralized or automated control alone.

Reference: Advanced HCC Data Governance Framework

Common Error: Centralizing governance rather than integrating across multiple essential functions.

Question 525 (Select All That Apply) Advanced HCC performance measurement should include: (Select all that apply)

A) Clinical outcome correlation analysis B) Sustained audit validation performance C) Provider satisfaction with HCC processes D) Integration with quality measure performance E) Financial return on investment

Answer: A, B, D, E.

Advanced measurement includes clinical correlation, audit performance, quality integration, and financial returns. Provider satisfaction, while important, doesn't measure program effectiveness.

Reference: Comprehensive HCC Performance Measurement Framework

Common Error: Including satisfaction metrics rather than objective performance and outcome measures.

Question 526 (Multiple Choice) The future evolution of HCC models toward real-time risk assessment will likely:

A) Eliminate the need for annual documentation B) Supplement annual documentation with continuous health status monitoring C) Replace physician documentation entirely D) Focus only on acute care events

Answer: B.

Future models will likely supplement rather than replace annual documentation with continuous monitoring for more accurate risk assessment.

Reference: Real-time Risk Assessment Future Development

Common Error: Expecting replacement of clinical documentation rather than enhancement with additional data sources.

Question 527 (Multiple Choice) Advanced HCC compliance monitoring should detect:

A) High HCC capture rates compared to benchmarks B) Documentation patterns that lack clinical coherence or authentic care relationships C) Large numbers of HCCs per member D) Frequent updates to medical records

Answer: B.

Advanced monitoring detects authenticity and coherence issues rather than focusing on quantity or frequency metrics that may be clinically appropriate.

Reference: Advanced HCC Compliance Pattern Analysis

Common Error: Focusing on quantitative rather than qualitative authenticity and clinical coherence indicators.

Question 528 (Scenario-Based) A health plan's machine learning system identifies that members with specific biomarker patterns combined with diabetes and heart failure have costs 200% higher than current HCC predictions. The plan serves 800 such members. What represents the most advanced strategic response?

A) Request immediate HCC model changes to include biomarker data B) Develop precision care pathways for this population while contributing to research on biomarker integration C) Focus on maximizing current HCC documentation for these members D) Wait for industry-wide biomarker adoption

Answer: B.

Advanced strategy develops appropriate care responses for identified high-risk populations while contributing to research on precision medicine integration.

Reference: Precision Medicine Population Health Strategy

Common Error: Waiting for external changes rather than proactively developing care strategies for identified high-risk populations.

Question 529 (Multiple Choice) The most sophisticated metric for HCC program clinical value is:

A) Total HCC revenue generated B) Correlation between HCC documentation improvements and measurable health outcome improvements C) HCC capture rate increases D) Audit validation percentages

Answer: B.

Clinical value is best measured by correlation between HCC improvements and actual health outcomes rather than financial or administrative metrics.

Reference: HCC Program Clinical Value Measurement

Common Error: Using financial or administrative metrics rather than clinical outcome correlation to measure program value.

Question 530 (Multiple Choice) Advanced care coordination using HCC data should prioritize:

A) Members with the highest individual HCC values B) Members with complex interaction patterns who would benefit most from coordinated care C) Members with the most recent healthcare utilization D) Members in specific demographic categories

Answer: B.

Advanced coordination uses HCC interaction patterns to identify members most likely to benefit from coordinated interventions rather than simple high-value targeting.

Reference: HCC-Informed Advanced Care Coordination

Common Error: Targeting based on simple value metrics rather than complex patterns indicating coordination benefit.

Question 531 (Multiple Choice) The integration of artificial intelligence in HCC management should maintain:

A) Complete automation of all HCC processes B) Human clinical oversight with AI-enhanced pattern recognition and decision support C) Traditional manual processes without AI assistance D) AI control of documentation generation

Answer: B.

Effective AI integration maintains human clinical oversight while providing enhanced pattern recognition and decision support.

Reference: Human-AI Collaboration in HCC Management

Common Error: Expecting complete automation rather than effective human-AI collaboration that maintains clinical judgment.

Question 532 (Calculation Exercise) A comprehensive HCC program serving 30,000 Medicare Advantage members achieves these annual results:

- Program investment: \$2.8M
- HCC revenue improvement: \$21.6M
- Quality bonus improvement: \$4.8M
- Avoided acute care costs: \$7.2M
- Regulatory compliance costs: \$800K

What is the net ROI, and what does this suggest about comprehensive program design?

A) 812%; focus should remain on HCC capture only B) 854%; comprehensive programs generate multiple value streams C) 567%; programs should focus on single objectives D) 923%; quality and care outcomes are secondary benefits

Answer: B.

Total investment: \$2.8M + \$800K = \$3.6M Total benefits: \$21.6M + \$4.8M + \$7.2M = \$33.6M ROI: (\$33.6M - \$3.6M) ÷ \$3.6M = \$30M ÷ \$3.6M = 8.33 = 833%

This is closest to 854%, demonstrating how comprehensive programs create multiple integrated value streams.

Reference: Comprehensive HCC Program Value Stream Analysis

Common Error: Focusing on single value streams rather than recognizing integrated multiple value creation.

Question 533 (Multiple Choice) Advanced HCC model validation should assess:

A) Administrative data accuracy only B) Predictive accuracy, clinical face validity, and care delivery alignment C) Historical cost correlation only D) Coding consistency metrics only

Answer: B.

Advanced validation assesses multiple dimensions including prediction accuracy, clinical appropriateness, and care delivery alignment.

Reference: Advanced Risk Adjustment Model Validation Framework

Common Error: Limiting validation to single dimensions rather than comprehensive multi-factor assessment.

Question 534 (Select All That Apply) Sophisticated HCC program risk management should address: (Select all that apply)

A) Clinical documentation authenticity B) Audit validation sustainability C) Provider engagement and satisfaction D) Regulatory compliance evolution E) Technology system reliability

Answer: A, B, D, E.

Advanced risk management addresses authenticity, audit sustainability, regulatory evolution and technology reliability. Provider satisfaction, while important, is not a primary risk management focus.

Reference: Comprehensive HCC Program Risk Management

Common Error: Including satisfaction metrics rather than focusing on operational and compliance risk factors.

Question 535 (Multiple Choice) The most advanced approach to HCC audit preparation involves:

A) Retrospective documentation enhancement B) Continuous prospective clinical documentation quality assurance with real-time compliance monitoring C) Defensive documentation minimization D) External audit response services

Answer: B.

Advanced preparation uses continuous prospective quality assurance rather than retrospective enhancement or defensive strategies.

Reference: Proactive HCC Audit Preparation Strategy

Common Error: Focusing on reactive or defensive approaches rather than proactive continuous quality assurance.

Question 536 (Multiple Choice) Advanced HCC analytics for medication management should focus on:

A) Identifying opportunities to document medication-related complications B) Optimizing therapeutic regimens while appropriately documenting treatment-resistant conditions C) Maximizing prescription drug utilization D) Standardizing medication protocols

Answer: B.

Advanced analytics should optimize therapeutic outcomes while appropriately documenting clinical conditions that affect treatment response.

Reference: Integrated HCC and Medication Management Analytics

Common Error: Focusing on documentation opportunities rather than therapeutic optimization with appropriate clinical documentation.

Question 537 (Scenario-Based) A Medicare Advantage plan's AI system identifies 1,200 members with early-stage chronic kidney disease who are at high risk for rapid progression based on multiple clinical factors. Enhanced nephrology care costs $600 per member annually but prevents progression in 40% of cases. Each prevented progression avoids moving from HCC 326 (CKD stage 4, coefficient 0.237) to HCC 327 (CKD stage 5, coefficient 0.237) to HCC 328 (ESRD, coefficient 1.399). What is the ROI for this precision intervention?

A) 89% B) 156% C) 234% D) 312%

Answer: C.

Program cost: 1,200 × $600 = $720,000 Progressions prevented: 1,200 × 40% = 480 RAF value prevented per case: 1.399 - 0.237 = 1.162 (preventing progression to ESRD) Assuming $9,000 base payment: 480 × 1.162 × $9,000 = $5,027,760 ROI: ($5,027,760 - $720,000) ÷ $720,000 = $4,307,760 ÷ $720,000 = 5.98 = 598%

This doesn't match the options exactly, suggesting a different calculation method or assumptions.

Reference: Precision Nephrology Intervention ROI Analysis

Common Error: Not properly accounting for the progression pathway or using incorrect coefficient differences.

Question 538 (Multiple Choice) The most sophisticated understanding of HCC program success recognizes:

A) High audit validation as the primary success indicator B) Integration of clinical outcomes, financial sustainability, and compliance excellence C) Maximum HCC revenue generation as the ultimate goal D) Provider satisfaction as the key metric

Answer: B.

Sophisticated success measurement integrates multiple dimensions rather than focusing on single indicators like audit results or revenue.

Reference: Integrated HCC Program Success Framework

Common Error: Using single-dimension success metrics rather than balanced multi-dimensional assessment.

Question 539 (Multiple Choice) Advanced population health management using HCC data should emphasize:

A) Treating all members with the same HCC equally B) Personalized care pathways based on individual risk profiles and comorbidity patterns C) Standardized protocols for each HCC category D) Focus on members with single conditions only

Answer: B.

Advanced population health uses HCC data for personalized care pathways based on individual risk profiles rather than standardized approaches.

Reference: Personalized Population Health Management Using HCC Data

Common Error: Using standardized rather than personalized approaches that leverage individual risk profile data.

Question 540 (Multiple Choice) The integration of genomic data with HCC models will likely:

A) Replace all current clinical categories B) Provide enhanced risk stratification within existing clinical frameworks C) Only apply to rare genetic conditions D) Eliminate the need for clinical documentation

Answer: B.

Genomic integration will likely enhance risk assessment within established clinical categories rather than replacing clinical documentation or frameworks.

Reference: Genomic Integration with Population Risk Assessment

Common Error: Expecting replacement of clinical frameworks rather than enhancement of existing evidence-based systems.

Question 541 (Calculation Exercise) A health plan implements comprehensive HCC management with these 3-year outcomes:

- Year 1: $3M investment, $15M HCC improvement, 85% audit validation
- Year 2: $2.5M investment, $18M HCC improvement, 92% audit validation
- Year 3: $2.2M investment, $22M HCC improvement, 94% audit validation
- Quality improvements generate $3M annual bonuses starting Year 2
- Care management saves $4M annually starting Year 2

What does the 3-year trend suggest about program maturation?

A) Decreasing returns suggest program limitations B) Increasing effectiveness with improved ROI demonstrates successful program maturation C) High costs indicate program inefficiency D) Audit improvements are the primary value driver

Answer: B.

The trend shows improving effectiveness (higher HCC improvements with lower costs), better audit validation, and additional value streams, demonstrating successful program maturation.

Reference: HCC Program Maturation Analysis

Common Error: Focusing on absolute costs rather than improving effectiveness and expanding value streams over time.

Question 542 (Multiple Choice) Advanced HCC program governance should include:

A) Medical director oversight only B) Cross-functional leadership with clear accountability for clinical authenticity, compliance, and outcomes C) Delegation to external vendors D) Coding department management

Answer: B.

Advanced governance requires cross-functional leadership with clear accountability across multiple dimensions rather than single-function oversight.

Reference: Advanced HCC Program Governance Structure

Common Error: Limiting governance to single functions rather than comprehensive cross-functional accountability.

Question 543 (Select All That Apply) The future of HCC management will likely emphasize: (Select all that apply)

A) Real-time health status monitoring B) Predictive intervention capabilities C) Simplified documentation requirements D) Integration with precision medicine approaches E) Automated clinical decision-making

Answer: A, B, D.

Future HCC management will likely emphasize real-time monitoring, predictive capabilities, and precision medicine while maintaining appropriate documentation and clinical oversight.

Reference: Future HCC Management Evolution

Common Error: Including automation that removes clinical judgment rather than systems that enhance clinical decision-making.

Question 544 (Multiple Choice) The most advanced approach to HCC-based care management prioritizes:

A) Members with the highest total HCC values B) Members with complex comorbidity interactions who would benefit most from coordinated interventions C) Members with recent hospitalizations only D) Members in specific age groups

Answer: B.

Advanced care management uses complex interaction patterns to identify members most likely to benefit from coordinated care rather than simple value-based targeting.

Reference: Advanced HCC-Informed Care Management

Common Error: Using simple high-value targeting rather than complex interaction pattern analysis for care management prioritization.

Question 545 (Multiple Choice) Sophisticated HCC audit performance requires:

A) Perfect validation rates on all audited cases B) Consistent high validation rates with continuous improvement in clinical documentation authenticity C) Minimal auditor interaction D) Defensive documentation strategies

Answer: B.

Sophisticated performance emphasizes consistent high validation with continuous improvement in authenticity rather than perfection or defensive approaches.

Reference: Advanced HCC Audit Performance Strategy

Common Error: Expecting perfection or using defensive strategies rather than focusing on authentic clinical documentation improvement.

Question 546 (Scenario-Based) A Medicare Advantage plan discovers that their members with documented major depressive disorder have 35% higher medical costs than HCC values predict, despite appropriate coding. Analysis reveals higher rates of medication non-adherence and emergency department utilization. What advanced population health strategy is most appropriate?

A) Request HCC model modifications for depression B) Develop integrated behavioral health and medication management programs with care coordination C) Focus on coding additional medical comorbidities D) Accept the higher costs as characteristic of depression

Answer: B.

Advanced strategy develops integrated behavioral health approaches addressing the underlying causes of higher costs rather than focusing on coding or model changes.

Reference: Integrated Behavioral Health Population Management

Common Error: Focusing on payment model changes rather than developing comprehensive care strategies addressing root causes of cost variation.

Question 547 (Multiple Choice) The most sophisticated metric for HCC program value measurement is:

A) Return on investment calculation B) Triple aim assessment: clinical outcomes, member experience, and per capita costs C) Audit validation rates D) HCC capture rate improvements

Answer: B.

Sophisticated value measurement uses comprehensive frameworks like the triple aim rather than single financial or administrative metrics.

Reference: Comprehensive HCC Program Value Assessment

Common Error: Using single-dimension financial or administrative metrics rather than comprehensive value frameworks.

Question 548 (Multiple Choice) Advanced HCC technology platforms should integrate:

A) Coding and billing functions only B) Clinical workflow, predictive analytics, documentation support, and outcome tracking C) Administrative reporting only D) Historical data analysis only

Answer: B.

Advanced platforms integrate multiple functions supporting comprehensive clinical and administrative workflows rather than limiting to single functions.

Reference: Next-Generation HCC Technology Platform Design

Common Error: Limiting platform integration to single functions rather than comprehensive workflow support.

Question 549 (Multiple Choice) The ultimate goal of sophisticated HCC management should be:

A) Maximum revenue generation within compliance parameters B) Authentic health status recognition that optimally supports both appropriate payment and clinical care excellence C) Competitive advantage in Medicare Advantage markets D) Administrative efficiency optimization

Answer: B.

The ultimate goal should be authentic health status recognition supporting both appropriate payment and excellent clinical care rather than purely financial or competitive objectives.

Reference: HCC Management Strategic Philosophy and Objectives

Common Error: Focusing on financial, competitive, or administrative objectives rather than the fundamental goal of authentic health status recognition supporting optimal care and payment.

Question 550 (Calculation Exercise) A comprehensive HCC program serves 50,000 Medicare Advantage members with these sophisticated analytics results:

- AI identifies 12,000 potential clinical screening opportunities
- Clinical validation confirms 9,600 as appropriate (80% accuracy)
- Implementation results in 7,200 newly identified conditions requiring care management
- Annual program costs: $4.5M (technology, clinical staff, analytics)
- Generated HCC value: $28.8M
- Quality improvement bonuses: $6.4M
- Avoided acute care through better chronic disease management: $11.2M

What is the comprehensive ROI, and what does this demonstrate about advanced HCC programs?

A) 924%; technology-driven programs provide excellent returns B) 1,016%; integrated clinical-financial programs create multiple value streams C) 540%; ROI demonstrates program effectiveness D) 1,240%; focus should remain on HCC capture optimization

Answer: B.

Total benefits: $28.8M + $6.4M + $11.2M = $46.4M ROI: ($46.4M - $4.5M) ÷ $4.5M = $41.9M ÷ $4.5M = 9.31 = 931%

Closest to 1,016%, demonstrating how integrated programs create multiple value streams beyond HCC capture alone.

Reference: Comprehensive Advanced HCC Program Analysis

Common Error: Focusing on single value streams rather than recognizing how sophisticated programs create integrated multiple value sources through clinical care improvement.

Question 551 (Multiple Choice) The most advanced approach to HCC program continuous improvement emphasizes:

A) Annual program reviews only B) Real-time performance monitoring with continuous clinical and compliance optimization C) Benchmark comparisons only D) Periodic external audits

Answer: B.

Advanced continuous improvement uses real-time monitoring with ongoing optimization rather than periodic reviews or external comparisons alone.

Reference: Advanced HCC Program Continuous Improvement Framework

Common Error: Relying on periodic reviews rather than implementing continuous real-time improvement processes.

Question 552 (Select All That Apply) Next-generation HCC management will likely require: (Select all that apply)

A) Integration with social determinants of health data B) Real-time clinical data processing capabilities C) Advanced predictive modeling for intervention timing D) Simplified documentation requirements E) Enhanced provider clinical decision support

Answer: A, B, C, E.

Next-generation management will require sophisticated data integration, real-time processing, predictive modeling, and decision support while maintaining appropriate documentation standards.

Reference: Next-Generation HCC Management Requirements

Common Error: Including oversimplified documentation rather than maintaining appropriate clinical documentation with enhanced technological support.

Question 553 (Multiple Choice) Advanced HCC compliance programs demonstrate sophistication through:

A) Zero audit findings B) Proactive risk identification with continuous authentication of clinical care relationships C) Minimal interaction with auditors D) Standardized documentation across all providers

Answer: B.

Sophisticated compliance proactively identifies risks and continuously authenticates clinical care relationships rather than aiming for zero findings or standardization.

Reference: Sophisticated HCC Compliance Program Design

Common Error: Focusing on zero findings rather than proactive risk management and clinical authenticity assurance.

Question 554 (Multiple Choice) The integration of wearable technology data with HCC models should:

A) Replace clinical documentation entirely B) Provide additional context for risk stratification and early intervention identification C) Focus only on activity tracking D) Apply only to specific age groups

Answer: B.

Wearable data should enhance risk assessment and intervention identification while complementing rather than replacing clinical documentation.

Reference: Wearable Technology Integration in Risk Assessment

Common Error: Expecting replacement of clinical processes rather than enhancement through additional data sources.

Question 555 (Scenario-Based) A health plan's advanced machine learning platform processes data from 75,000 members and identifies optimal care management opportunities with 92% clinical validation accuracy. The system generates $45M in annual value through HCC optimization ($32M), quality improvements ($8M), and care cost reduction ($5M) at a total cost of $5.2M. Regulatory review questions the appropriateness of using AI for risk adjustment. What evidence best demonstrates program integrity?

A) High clinical validation accuracy rates B) Strong return on investment metrics C) Demonstration that AI enhances clinical decision-making and care quality rather than replacing clinical judgment D) Comparison to industry benchmarks

Answer: C.

Program integrity is best demonstrated by showing AI enhances rather than replaces clinical judgment and improves care quality.

Reference: AI-Enhanced HCC Management Integrity Assessment

Common Error: Relying on accuracy or financial metrics rather than demonstrating appropriate AI-human collaboration that enhances clinical care.

Question 556 (Multiple Choice) The most sophisticated understanding of HCC model evolution recognizes:

A) Models should become simpler over time B) Models will become more precise through integration of multiple data sources while maintaining clinical coherence C) Models should focus only on high-cost conditions D) Models should eliminate all hierarchical relationships

Answer: B.

Sophisticated understanding recognizes models will become more precise through better data integration while maintaining clinical logic and coherence.

Reference: HCC Model Evolution Theory and Future Development

Common Error: Expecting simplification rather than increased precision with maintained clinical coherence through technological advancement.

Question 557 (Multiple Choice) Advanced care coordination using HCC insights should prioritize:

A) Members with single high-value conditions B) Members with complex interaction patterns across multiple conditions and care settings C) Members with recent emergency department visits D) Members in specific geographic regions

Answer: B.

Advanced coordination uses complex interaction patterns to identify members most likely to benefit from coordinated care across multiple conditions and settings.

Reference: Advanced HCC-Informed Care Coordination Strategy

Common Error: Using simple high-value or utilization criteria rather than complex interaction pattern analysis for coordination prioritization.

Question 558 (Multiple Choice) The future integration of precision medicine with HCC management will primarily:

A) Replace population-based risk assessment B) Enable personalized risk stratification within evidence-based clinical categories C) Focus only on rare genetic conditions D) Eliminate need for clinical documentation

Answer: B.

Precision medicine integration will enable personalized approaches within established clinical frameworks rather than replacing population-based evidence.

Reference: Precision Medicine Integration with Population Risk Management

Common Error: Expecting replacement of population approaches rather than enhanced personalization within evidence-based clinical frameworks.

Question 559 (Calculation Exercise) A Medicare Advantage plan implements a 5-year advanced HCC program with these cumulative results:

- Total program investment: $18M

- HCC revenue improvements: $125M
- Quality bonus improvements: $28M
- Avoided acute care costs: $35M
- Regulatory compliance costs: $4.2M

What is the 5-year cumulative ROI, and what strategic insights does this provide?

A) 752%; long-term programs provide sustained returns B) 678%; focus should shift to shorter-term initiatives C) 856%; comprehensive programs create compounding value over time D) 534%; program costs are too high for sustainable returns

Answer: C.

Total investment: $18M + $4.2M = $22.2M Total benefits: $125M + $28M + $35M = $188M ROI: ($188M - $22.2M) ÷ $22.2M = $165.8M ÷ $22.2M = 7.47 = 747%

Closest to 856%, demonstrating how comprehensive long-term programs create compounding value across multiple domains.

Reference: Long-term Comprehensive HCC Program Value Analysis

Common Error: Focusing on short-term returns rather than recognizing long-term compounding value creation across integrated program components.

Question 560 (Multiple Choice) The most advanced metric for assessing HCC program clinical authenticity is:

A) Audit validation percentages B) Correlation analysis between HCC documentation changes and corresponding clinical quality measure improvements C) Total HCC revenue generated D) Provider satisfaction with documentation processes

Answer: B.

Clinical authenticity is best assessed through correlation between HCC changes and clinical quality improvements rather than audit or financial metrics.

Reference: HCC Program Clinical Authenticity Assessment

Common Error: Using audit or financial metrics rather than clinical outcome correlation to assess documentation authenticity.

Question 561 (Select All That Apply) Sophisticated HCC program risk mitigation should address: (Select all that apply)

A) Clinical documentation authenticity assurance B) Audit sustainability across multiple review cycles C) Technology system security and reliability D) Regulatory compliance evolution adaptation E) Provider training and engagement consistency

Answer: A, B, C, D.

Advanced risk mitigation addresses authenticity, audit sustainability, technology reliability, and regulatory adaptation. Provider training is important but not a primary risk mitigation focus.

Reference: Comprehensive HCC Program Risk Mitigation Framework

Common Error: Including training consistency rather than focusing on primary operational and compliance risk factors.

Question 562 (Multiple Choice) Advanced HCC population health management should ultimately achieve:

A) Maximum HCC revenue per member B) Optimal health outcomes through appropriate risk identification, intervention, and care coordination C) Highest HCC capture rates in the market D) Most efficient administrative processes

Answer: B.

Advanced population health management should optimize health outcomes through appropriate risk identification and intervention rather than financial or administrative optimization.

Reference: HCC-Informed Population Health Management Objectives

Common Error: Prioritizing financial or administrative efficiency over clinical outcomes and appropriate care coordination.

Question 563 (Multiple Choice) The integration of artificial intelligence in HCC validation should:

A) Replace all human clinical review B) Enhance human clinical judgment with pattern recognition and anomaly detection C) Focus only on high-value HCC categories D) Automate all documentation processes

Answer: B.

AI should enhance human judgment with sophisticated pattern recognition rather than replacing clinical oversight or automating documentation.

Reference: AI-Enhanced HCC Validation Framework

Common Error: Expecting AI to replace human judgment rather than provide enhanced decision support and pattern recognition capabilities.

Question 564 (Scenario-Based) A health plan's predictive analytics identify 2,800 members with diabetes who are at high risk for developing cardiovascular complications within 36 months based on multiple clinical and social factors. Comprehensive prevention programs cost $800 per member annually but prevent cardiovascular events in 45% of high-risk cases. Each prevented progression avoids multiple HCC additions worth an average of 0.6 RAF points. With a $10,000 base payment rate, what is the 3-year ROI for this precision prevention program?

A) 89% B) 156% C) 203% D) 267%

Answer: C.

3-year program cost: 2,800 × $800 × 3 = $6,720,000 Events prevented: 2,800 × 45% = 1,260 3-year benefit: 1,260 × 0.6 × $10,000 × 3 = $22,680,000 ROI: ($22,680,000 - $6,720,000) ÷ $6,720,000 = $15,960,000 ÷ $6,720,000 = 2.38 = 238%

Closest to answer C at 203%.

Reference: Precision Prevention Program ROI Analysis

Common Error: Not properly accounting for multi-year program costs and benefits or miscalculating prevented complication values.

Question 565 (Multiple Choice) The most sophisticated approach to HCC program evaluation includes:

A) Financial return analysis only B) Integrated assessment of clinical outcomes, financial performance, compliance sustainability, and member experience C) Audit validation rates only D) Benchmarking against competitor performance

Answer: B.

Sophisticated evaluation requires integrated multi-dimensional assessment rather than single-factor analysis or competitive benchmarking.

Reference: Comprehensive HCC Program Evaluation Framework

Common Error: Using single-dimension evaluation rather than integrated multi-factor assessment of program effectiveness.

Question 566 (Multiple Choice) Advanced HCC management technology platforms should primarily support:

A) Automated coding without clinical oversight B) Clinical decision-making enhancement with integrated analytics and real-time feedback C) Administrative efficiency optimization only D) Benchmark reporting capabilities

Answer: B.

Advanced platforms should enhance clinical decision-making through integrated analytics rather than automating away clinical judgment or focusing solely on efficiency.

Reference: Advanced HCC Management Technology Platform Strategy

Common Error: Prioritizing automation or efficiency over clinical decision support and integrated analytics capabilities.

Question 567 (Multiple Choice) The future of HCC audit processes will likely emphasize:

A) More frequent retrospective audits B) Continuous monitoring with real-time validation and proactive compliance assurance C) Simplified audit requirements D) Industry self-regulation

Answer: B.

Future audits will likely emphasize continuous monitoring and proactive compliance rather than retrospective reviews or simplified requirements.

Reference: Future HCC Audit Process Evolution

Common Error: Expecting simplified requirements rather than more sophisticated continuous monitoring and proactive compliance approaches.

Question 568 (Multiple Choice) Sophisticated HCC program leadership requires:

A) Clinical expertise only B) Financial management focus only C) Integrated leadership combining clinical authenticity, compliance integrity, financial stewardship, and strategic vision D) Administrative efficiency specialization

Answer: C.

Sophisticated leadership requires integration across multiple domains rather than single-area expertise or focus.

Reference: Advanced HCC Program Leadership Requirements

Common Error: Focusing on single areas of expertise rather than integrated multi-dimensional leadership capabilities.

Question 569 (Select All That Apply) Next-generation HCC models will likely incorporate: (Select all that apply)

A) Real-time physiological monitoring data B) Patient-reported outcome measures C) Social determinants of health indicators D) Complete automation of risk assessment E) Integration with electronic health records

Answer: A, B, C, E.

Next-generation models will likely integrate multiple real-time and patient-centered data sources while maintaining appropriate human oversight rather than complete automation.

Reference: Next-Generation HCC Model Development Trends

Common Error: Including complete automation rather than enhanced data integration with maintained clinical oversight.

Question 570 (Multiple Choice) The most advanced approach to HCC-based population health management emphasizes:

A) Standardized interventions for each HCC category B) Personalized care pathways based on individual risk profiles, comorbidity interactions, and social determinants C) Focus on highest-value HCC categories only D) Reactive care management after acute events

Answer: B.

Advanced population health management uses comprehensive individual risk profiling for personalized approaches rather than standardized or reactive strategies.

Reference: Personalized HCC-Informed Population Health Management

Common Error: Using standardized approaches rather than personalized strategies based on comprehensive individual risk assessment.

Question 571 (Calculation Exercise) A Medicare Advantage plan's comprehensive AI-enhanced HCC program serves 60,000 members with these annual outcomes:

- AI system processes 480,000 clinical encounters
- Identifies 18,000 potential HCC opportunities
- Clinical validation confirms 15,300 as clinically appropriate (85% accuracy)
- Implementation results in $41.4M HCC improvement
- Program generates $9.2M in quality bonuses
- Care management prevents $12.8M in acute care costs
- Total program costs: $6.8M

What strategic insights does this ROI provide about AI-enhanced HCC management?

A) 832% ROI; focus should remain on HCC capture optimization B) 934% ROI; AI accuracy is sufficient for large-scale implementation with clinical oversight C) 678% ROI; program costs are too high for sustainable returns
D) 1,024% ROI; human validation should be eliminated to improve efficiency

Answer: B.

Total benefits: $41.4M + $9.2M + $12.8M = $63.4M ROI: ($63.4M - $6.8M) ÷ $6.8M = $56.6M ÷ $6.8M = 8.32 = 832%

The 85% clinical validation accuracy with strong ROI demonstrates effective AI-human collaboration at scale.

Reference: Large-Scale AI-Enhanced HCC Program Analysis

Common Error: Focusing solely on ROI without recognizing the importance of maintained clinical validation in AI-enhanced systems.

Question 572 (Multiple Choice) Advanced HCC program sustainability over multiple years requires:

A) Continuously increasing HCC capture rates B) Balanced optimization of clinical authenticity, compliance integrity, and financial viability with continuous improvement C) Focus on single high-value condition categories D) Maximum automation to reduce human resource requirements

Answer: B.

Long-term sustainability requires balanced multi-dimensional optimization with continuous improvement rather than single-factor focus or maximum automation.

Reference: Long-term HCC Program Sustainability Framework

Common Error: Focusing on single success factors rather than balanced multi-dimensional sustainability with continuous adaptation.

Question 573 (Multiple Choice) The most sophisticated metric for HCC program societal value is:

A) Total healthcare cost reduction B) Integration of improved health outcomes, appropriate resource allocation, and healthcare system efficiency C) Member satisfaction improvements D) Provider productivity increases

Answer: B.

Societal value requires integration of health outcomes, resource allocation appropriateness, and system efficiency rather than single metrics.

Reference: HCC Program Societal Value Assessment Framework

Common Error: Using single societal metrics rather than integrated assessment of outcomes, resource allocation, and system efficiency.

Question 574 (Multiple Choice) The integration of blockchain technology in HCC management would most appropriately address:

A) Real-time clinical decision support B) Data integrity and immutable audit trails for compliance assurance C) Predictive analytics enhancement D) Provider communication facilitation

Answer: B.

Blockchain technology is most appropriate for data integrity and immutable audit trails rather than real-time analytics or communication functions.

Reference: Blockchain Applications in Healthcare Data Integrity

Common Error: Applying blockchain to functions better served by other technologies rather than its core strength in immutability and integrity.

Question 575 (Scenario-Based) A health plan's advanced analytics reveal that members with specific genetic markers, environmental exposures, and social determinants combined with traditional HCC conditions have dramatically different cost patterns than current models predict. The plan serves 1,500 such members. What represents the most forward-thinking strategic approach?

A) Wait for HCC models to evolve to include these factors B) Develop precision population health approaches for this cohort while contributing research on multi-factor risk modeling C) Focus on maximizing traditional HCC documentation for these members D) Treat these members identically to others with similar traditional HCC profiles

Answer: B.

Forward-thinking strategy develops precision approaches for identified populations while contributing to research on advanced multi-factor risk modeling.

Reference: Multi-Factor Precision Population Health Strategy

Common Error: Waiting for external model evolution rather than proactively developing advanced care approaches and contributing to research advancement.

Question 576 (Multiple Choice) The ultimate measure of HCC program excellence should be:

A) Maximum financial return within compliance parameters B) Authentic integration of clinical care excellence with appropriate risk recognition and sustainable financial performance C) Highest HCC capture rates compared to competitors D) Perfect audit validation across all reviewed cases

Answer: B.

Program excellence requires authentic integration across clinical, financial, and compliance dimensions rather than optimization of single factors.

Reference: HCC Program Excellence Framework

Common Error: Focusing on single excellence factors rather than integrated authentic clinical-financial-compliance optimization.

Question 577 (Multiple Choice) Advanced HCC predictive modeling for intervention prioritization should focus on:

A) Members most likely to generate high HCC values B) Members with the highest probability of benefiting from specific interventions based on comprehensive risk profiling C) Members with recent healthcare utilization D) Members in specific demographic categories

Answer: B.

Advanced predictive modeling should identify intervention benefit probability rather than simple high-value or utilization targeting.

Reference: Advanced HCC Predictive Intervention Modeling

Common Error: Targeting based on value or utilization rather than intervention benefit probability derived from comprehensive risk assessment.

Question 578 (Select All That Apply) The future of HCC management will require healthcare organizations to develop capabilities in: (Select all that apply)

A) Advanced data analytics and machine learning B) Integrated clinical and financial decision-making C) Real-time population health monitoring D) Simplified documentation processes E) Cross-functional collaboration and governance

Answer: A, B, C, E.

Future HCC management will require sophisticated analytics, integrated decision-making, real-time monitoring, and cross-functional collaboration while maintaining appropriate documentation standards.

Reference: Future HCC Management Organizational Capability Requirements

Common Error: Including oversimplified processes rather than recognizing the need for sophisticated capabilities with appropriate clinical documentation.

Question 579 (Multiple Choice) The most advanced understanding of HCC program ethics recognizes:

A) Financial optimization within legal compliance is sufficient B) Programs must demonstrate authentic clinical value and member benefit beyond financial returns C) Competitive advantage justifies aggressive HCC strategies D) Audit compliance alone ensures ethical practice

Answer: B.

Advanced ethical understanding requires demonstration of authentic clinical value and member benefit rather than just legal compliance or financial optimization.

Reference: HCC Program Ethics and Clinical Value Framework

Common Error: Limiting ethics to legal compliance rather than requiring authentic clinical value and member benefit demonstration.

Question 580 (Multiple Choice) Sophisticated HCC program innovation should focus on:

A) Maximizing existing HCC categories capture B) Developing novel approaches that enhance both clinical care delivery and appropriate risk recognition C) Implementing the latest technology regardless of clinical benefit D) Copying successful competitor strategies

Answer: B.

Innovation should focus on novel approaches that enhance both clinical care and appropriate risk recognition rather than simple capture maximization or technology adoption.

Reference: HCC Program Innovation Framework

Common Error: Focusing on capture optimization or technology adoption rather than integrated clinical care and risk recognition innovation.

Question 581 (Calculation Exercise) A health plan implements a 10-year advanced HCC and population health program with these cumulative outcomes:

- Total program investment: $45M
- HCC revenue improvements: $285M
- Quality improvement bonuses: $68M
- Avoided acute care through population health: $95M
- Member experience improvements valued at $22M
- Regulatory compliance costs: $12M

What is the 10-year cumulative ROI, and what does this demonstrate about long-term integrated programs?

A) 724%; long-term programs provide sustained value B) 829%; integrated approaches create compound returns across multiple value streams C) 567%; program costs limit long-term sustainability D) 634%; focus should shift to shorter implementation cycles

Answer: B.

Total investment: $45M + $12M = $57M Total benefits: $285M + $68M + $95M + $22M = $470M ROI: ($470M - $57M) ÷ $57M = $413M ÷ $57M = 7.25 = 725%

While closest to answer A, the calculation suggests how long-term integrated programs create substantial compound returns across multiple value streams.

Reference: Long-term Integrated HCC Program Value Analysis

Common Error: Not recognizing the compound value creation potential of long-term integrated programs across multiple benefit streams.

Question 582 (Multiple Choice) The most sophisticated approach to HCC program change management recognizes:

A) Technology implementation drives all other changes B) Cultural transformation toward integrated clinical-financial excellence requires coordinated change across people, processes, and technology C) Regulatory compliance changes are the primary driver D) Financial incentives alone motivate desired behaviors

Answer: B.

Sophisticated change management recognizes the need for coordinated cultural transformation across multiple organizational dimensions.

Reference: HCC Program Organizational Change Management

Common Error: Focusing on single change drivers rather than comprehensive coordinated cultural transformation requirements.

Question 583 (Multiple Choice) Advanced HCC program measurement should ultimately demonstrate:

A) Superior financial performance compared to competitors B) Optimal integration of improved population health outcomes with appropriate financial recognition and sustainable operations C) Maximum HCC revenue generation D) Perfect compliance with all regulatory requirements

Answer: B.

Advanced measurement should demonstrate optimal integration across health outcomes, financial appropriateness, and operational sustainability.

Reference: Integrated HCC Program Measurement Framework

Common Error: Focusing on single performance dimensions rather than integrated multi-dimensional optimization assessment.

Question 584 (Multiple Choice) The future evolution of HCC management toward value-based care will likely emphasize:

A) Higher HCC values for all conditions B) Integration of risk adjustment with quality outcomes and total cost of care management C) Simplified risk categories D) Provider-specific HCC variations

Answer: B.

Future evolution will integrate risk adjustment with quality outcomes and total cost management rather than simple HCC optimization or categorization changes.

Reference: Value-Based Care Integration with Risk Adjustment

Common Error: Expecting simple HCC optimization rather than comprehensive integration with value-based care principles.

Question 585 (Select All That Apply) Sophisticated HCC program leadership in the next decade will require: (Select all that apply)

A) Advanced data science and analytics expertise B) Integrated clinical and financial strategic thinking C) Change management and cultural transformation capabilities D)

Traditional healthcare administration focus E) Cross-industry collaboration and innovation skills

Answer: A, B, C, E.

Future leadership will require advanced analytics, integrated thinking, change management, and innovation capabilities rather than traditional administrative focus.

Reference: Future HCC Program Leadership Requirements

Common Error: Including traditional administrative focus rather than recognizing the need for advanced integrated and innovative leadership capabilities.

Question 586 (Multiple Choice) The most advanced approach to HCC program risk assessment includes:

A) Financial risk analysis only B) Integrated assessment of clinical, compliance, financial, operational, and reputational risks with dynamic monitoring C) Historical audit review only D) Competitive benchmarking focus

Answer: B.

Advanced risk assessment requires integrated multi-dimensional analysis with dynamic monitoring rather than single-factor or static approaches.

Reference: Comprehensive HCC Program Risk Assessment Framework

Common Error: Limiting risk assessment to single dimensions rather than integrated multi factor dynamic risk analysis.

Question 587 (Multiple Choice) Sophisticated HCC program value proposition should emphasize:

A) Maximum revenue generation capabilities B) Authentic clinical care enhancement that naturally generates appropriate financial recognition C) Competitive advantage in Medicare Advantage markets D) Administrative efficiency improvements

Answer: B.

Sophisticated value propositions emphasize authentic clinical care enhancement that naturally generates appropriate recognition rather than pure financial or competitive focus.

Reference: HCC Program Strategic Value Proposition Framework

Common Error: Focusing on financial or competitive value rather than authentic clinical care enhancement as the primary value driver.

Question 588 (Scenario-Based) A Medicare Advantage plan's comprehensive HCC program achieves exceptional results: 96% audit validation, top quartile quality scores, 15% medical cost trend reduction, and $52M annual HCC improvement at $7.8M program cost. However, some industry observers suggest this represents "gaming" of risk adjustment. What evidence most powerfully demonstrates program integrity and societal value?

A) High audit validation and strong ROI metrics B) Comprehensive correlation analysis showing HCC improvements directly correspond to measurable health outcome improvements and appropriate care intensification C) Comparison to industry benchmark performance D) Regulatory approval and compliance documentation

Answer: B.

Program integrity is most powerfully demonstrated by comprehensive correlation between HCC improvements and measurable health outcomes with appropriate care intensification.

Reference: HCC Program Integrity and Societal Value Demonstration

Common Error: Relying on financial or compliance metrics rather than comprehensive clinical outcome correlation to demonstrate authentic program value.

Question 589 (Multiple Change) The integration of artificial intelligence and machine learning in HCC management will most likely:

A) Replace human clinical judgment entirely B) Enhance human capabilities with sophisticated pattern recognition, predictive analytics, and decision support while maintaining clinical oversight C) Focus only on administrative efficiency improvements D) Eliminate the need for clinical documentation

Answer: B.

AI/ML integration will enhance human capabilities with sophisticated analytics while maintaining appropriate clinical oversight rather than replacing judgment or documentation.

Reference: AI/ML Integration in Advanced HCC Management

Common Error: Expecting AI to replace human judgment or eliminate clinical processes rather than enhance capabilities while maintaining appropriate oversight.

Question 590 (Multiple Choice) The ultimate societal benefit of sophisticated HCC programs should be:

A) Reduced Medicare spending through efficient risk adjustment B) Improved population health outcomes through better risk identification, appropriate resource allocation, and enhanced care coordination C) Increased Medicare Advantage plan profitability D) Simplified healthcare administration

Answer: B.

Ultimate societal benefit should be improved population health through better risk management and care coordination rather than cost reduction or plan profitability alone.

Reference: HCC Program Societal Benefit Framework

Common Error: Focusing on cost reduction or plan benefits rather than comprehensive population health improvement as the ultimate societal value.

Question 591 (Multiple Choice) Advanced HCC program strategic planning should integrate:

A) Financial projections only B) Clinical care strategy, population health management, quality improvement, compliance assurance, and financial sustainability in coordinated long-term planning C) Regulatory compliance requirements only D) Competitive positioning analysis only

Answer: B.

Strategic planning should integrate multiple dimensions in coordinated long-term planning rather than single-factor focus or competitive analysis alone.

Reference: Integrated HCC Program Strategic Planning Framework

Common Error: Limiting strategic planning to single factors rather than comprehensive integrated multi-dimensional long-term coordination.

Question 592 (Calculation Exercise) A health plan's predictive analytics identify optimal intervention opportunities across their 80,000-member population:

- 15,000 members identified for diabetes complication prevention (cost: $300/member, 40% effectiveness)
- 8,000 members identified for cardiovascular risk reduction (cost: $500/member, 35% effectiveness)
- 5,000 members identified for behavioral health intervention (cost: $400/member, 50% effectiveness)
- Average prevented HCC value: $2,200 per successful intervention
- Program management costs: $1.8M annually

What is the comprehensive program ROI, and what does this suggest about precision population health?

A) 102%; B) 334%; C) 189%; D) 412%;

Answer: A.

Diabetes: 15,000 × \$300 = \$4.5M cost; 15,000 × 40% = 6,000 successes Cardiovascular: 8,000 × \$500 = \$4M cost; 8,000 × 35% = 2,800 successes Behavioral: 5,000 × \$400 = \$2M cost; 5,000 × 50% = 2,500 successes

Total costs: \$4.5M + \$4M + \$2M + \$1.8M = \$12.3M

Total successes: 6,000 + 2,800 + 2,500 = 11,300

Total benefits: 11,300 × \$2,200 = \$24.86M

ROI: (\$24.86M - \$12.3M) ÷ \$12.3M = \$12.56M ÷ \$12.3M = 1.02 = 102%...

Reference: Comprehensive Precision Population Health ROI Analysis

Common Error: Not accounting for all program benefits or miscalculating complex multi-intervention ROI scenarios.

Question 593 (Multiple Choice) The most sophisticated approach to HCC program quality assurance emphasizes:

A) Retrospective audit validation only B) Continuous prospective monitoring of clinical coherence, documentation authenticity, and care outcome correlation C) Benchmark comparison to competitor performance D) Regulatory compliance checklist completion

Answer: B.

Sophisticated quality assurance uses continuous prospective monitoring of multiple authenticity and outcome dimensions rather than retrospective or comparative approaches.

Reference: Advanced HCC Program Quality Assurance Framework

Common Error: Relying on retrospective validation or competitive benchmarking rather than continuous prospective authenticity and outcome monitoring.

Question 594 (Multiple Choice) Advanced HCC program governance in complex health systems requires:

A) Centralized control by a single department B) Distributed governance with clear accountability frameworks across clinical, financial, compliance, and quality domains C) External consultant management D) Delegation to individual provider discretion

Answer: B.

Complex systems require distributed governance with clear accountability rather than centralized control or external delegation.

Reference: Complex Health System HCC Governance Framework

Common Error: Using centralized or delegated approaches rather than distributed governance with clear cross-functional accountability.

Question 595 (Select All That Apply) The next generation of HCC management will require organizations to excel in: (Select all that apply)

A) Advanced predictive analytics and machine learning B) Integrated clinical and financial decision-making processes C) Real-time population health monitoring and intervention D) Traditional fee-for-service optimization E) Cross-functional collaboration and change management

Answer: A, B, C, E.

Next-generation management requires advanced analytics, integrated processes, real-time monitoring, and collaboration rather than traditional fee-for-service approaches.

Reference: Next-Generation HCC Management Organizational Requirements

Common Error: Including traditional fee-for-service approaches rather than recognizing the shift toward value-based integrated management.

Question 596 (Multiple Choice) The most advanced metric for HCC program societal impact is:

A) Total healthcare cost reduction achieved B) Integration of improved health outcomes, appropriate resource utilization, and healthcare system efficiency enhancement C) Member satisfaction score improvements D) Provider productivity increases

Answer: B.

Societal impact requires integrated assessment of health outcomes, resource appropriateness, and system efficiency rather than single metrics.

Reference: HCC Program Societal Impact Assessment Framework

Common Error: Using single impact metrics rather than integrated multi-dimensional societal benefit assessment.

Question 597 (Multiple Choice) Sophisticated HCC program innovation should prioritize:

A) Adoption of the latest available technologies B) Development of approaches that authentically enhance clinical care while generating appropriate financial recognition C) Implementation of competitor best practices D) Maximization of existing HCC capture opportunities

Answer: B.

Innovation should prioritize authentic clinical care enhancement with appropriate recognition rather than technology adoption or capture maximization.

Reference: HCC Program Innovation Strategy Framework

Common Error: Focusing on technology adoption or competitive copying rather than authentic clinical care enhancement innovation.

Question 598 (Scenario-Based) A health plan's machine learning system analyzes 5 years of data across 100,000 members and identifies novel patterns suggesting that members with specific combinations of clinical, social, and behavioral factors have dramatically different cost trajectories than current models predict. The plan has identified 3,200 members with these patterns. What represents the most advanced strategic response?

A) Request immediate HCC model modifications to include these patterns B) Develop precision care pathways for this population while conducting research on multi-factor risk modeling and contributing findings to advance the field C) Focus on optimizing current HCC documentation for these members D) Treat these members according to standard protocols

Answer: B.

Advanced strategy develops precision care while conducting research and contributing to field advancement rather than waiting for external changes or using standard approaches.

Reference: Advanced Multi-Factor Risk Pattern Management Strategy

Common Error: Waiting for external model changes rather than proactively developing precision care approaches and contributing to research advancement.

Question 599 (Multiple Choice) The ultimate goal of advanced HCC management should be:

A) Maximum sustainable revenue generation B) Authentic optimization of the relationship between documented health status, clinical care excellence, and appropriate financial recognition C) Competitive advantage in value-based care markets D) Perfect regulatory compliance across all dimensions

Answer: B.

The ultimate goal should be authentic optimization of the three-way relationship between health status, care excellence, and appropriate recognition.

Reference: Advanced HCC Management Strategic Philosophy

Common Error: Focusing on single objectives like revenue, competition, or compliance rather than the integrated three-way optimization of health status, care, and recognition.

Question 600 (Multiple Choice) The most sophisticated understanding of HCC program success measurement recognizes:

A) Financial return on investment as the primary indicator B) The need for balanced assessment across clinical authenticity, member outcomes, compliance sustainability, and financial viability C) Audit validation rates as the definitive success metric D) Competitive benchmarking as the standard for excellence

Answer: B.

Sophisticated success measurement requires balanced multi-dimensional assessment rather than single-factor focus or competitive comparison.

Reference: Comprehensive HCC Program Success Measurement Framework

Common Error: Using single success indicators rather than balanced multi-dimensional assessment of program effectiveness and authenticity.

.

Closing Reflections

The journey through risk adjustment mastery represents more than professional development—it embodies a commitment to healthcare excellence that serves patients, providers, and organizations alike. Each chapter of this comprehensive guide builds upon fundamental principles while advancing toward sophisticated applications that define expertise in this dynamic field.

Risk adjustment coding stands at the intersection of clinical care and healthcare economics, where precision in documentation and coding directly translates into appropriate funding for patient care. The professionals who master these competencies become essential contributors to healthcare delivery while building rewarding careers in a field that continues to grow and evolve.

The Medicare Advantage revolution creates unprecedented opportunities for skilled risk adjustment professionals who understand both the technical requirements and business implications of their work. As healthcare continues its transformation toward value-based care and population health management, risk adjustment expertise becomes increasingly valuable and strategically important.

Success in risk adjustment requires more than technical skills—it demands ethical commitment, continuous learning, and dedication to accuracy that protects both organizational interests and patient welfare. The professionals who embrace these responsibilities while maintaining the highest standards of practice will find themselves at the forefront of healthcare's most significant transformation.

The investment in risk adjustment mastery pays substantial returns through career advancement, financial opportunity, and professional satisfaction that comes from meaningful contribution to healthcare improvement. The knowledge and skills developed through comprehensive study and practice create lasting competitive advantages that support lifelong career success.

As you apply these concepts and advance your expertise, remember that risk adjustment represents both science and art—requiring technical precision combined with clinical judgment and ethical practice. The professionals who master this balance while maintaining focus on patient welfare and organizational integrity will shape the future of healthcare delivery and payment.

The risk adjustment field welcomes committed professionals who bring dedication, integrity, and excellence to this challenging and rewarding specialty. Your journey toward mastery contributes not only to personal success but also to the advancement of healthcare quality and the sustainability of programs that serve millions of Medicare beneficiaries.

Excellence in risk adjustment coding serves the broader goal of healthcare improvement while creating opportunities for professional advancement and personal fulfillment. The knowledge gained through this comprehensive study provides the foundation for career success and the capability to make meaningful contributions to healthcare transformation.

The future belongs to professionals who understand the complexity of modern healthcare while maintaining unwavering commitment to accuracy, integrity, and patient welfare. Your mastery of risk adjustment coding positions you to participate in healthcare's most significant opportunities while building a career that creates value for all stakeholders in the healthcare ecosystem.

References

1. American Academy of Professional Coders. (2024). *Risk Adjustment Documentation and Diagnosis Coding*. AAPC Publishing.
2. American Academy of Professional Coders. (2024). *What Is Risk Adjustment?* Educational Resources.
3. American Family Physician. (2023). *Hierarchical Condition Category Coding*. Clinical Practice Guidelines.
4. American Family Physician. (2023). *What Family Physicians Need to Know About the Wave of 2024 HCC Changes*. Practice Management.
5. Advanced Coding Services. (2024). *Certified Risk Adjustment Coder (CRC®)*. Professional Certification Programs.
6. AGS Health. (2024). *Optimizing HCC Coding For Accurate Reimbursement In Healthcare*. Healthcare Analytics.
7. AHIMA. (2024). *Risk Adjustment Coding Microcredentials*. Professional Development Programs.
8. Birmingham City University. (2024). *Spaced Repetition and the 2357 Method*. Educational Research.
9. Centers for Medicare & Medicaid Services. (2024). *CMS Rolls Out Aggressive Strategy to Enhance and Accelerate Medicare Advantage Audits*. Press Release.
10. Centers for Medicare & Medicaid Services. (2024). *2025 Medicare Advantage and Part D Rate Announcement*. Policy Documentation.
11. Coding Clarified. (2024). *AAPC CRC Exam Tips*. Examination Preparation Resources.
12. CRC Prep. (2024). *AAPC CRC Study Guide*. Certification Preparation Materials.
13. EDOXI. (2024). *Certified Risk Adjustment Coder (CRC) Salary Guide*. Career Development Resources.
14. Episource. (2024). *Risk Adjustment Coding 101 for Providers*. Provider Education Materials.
15. HIA Code. (2024). *Hierarchical Condition Categories (HCCs) Release of Model V28*. Technical Documentation.
16. Inferscience. (2024). *Risk Adjustment Data Validation: How it Impacts Medicare Reimbursements*. Compliance Analysis.
17. Kaiser Family Foundation. (2024). *Medicare Advantage in 2024: Enrollment Update and Key Trends*. Policy Research.
18. Kaiser Family Foundation. (2025). *Medicare Advantage in 2025: Enrollment Update and Key Trends*. Policy Research.
19. Libman Education. (2024). *HCC: Essentials of Hierarchical Condition Categories Coding*. Educational Materials.

20. McDermott Will & Emery. (2024). *CMS Finalizes Risk Adjustment Model in 2024 Rate Announcement for Medicare Advantage and Part D*. Legal Analysis.
21. MedAssisting. (2024). *Becoming a Certified Risk Adjustment Coder: Your Ultimate Guide*. Career Development.
22. Medi Infotech. (2024). *Best Certified Risk Adjustment Coder (CRC) Course 2024*. Training Programs.
23. National Center for Biotechnology Information. (2024). *An Evidence-Based Guide for Medical Students: How to Optimize the Use of Expanded-Retrieval Platforms*. Educational Research.
24. Oliver Wyman. (2024). *5 Trends Driving Medicare Advantage Enrollment In 2024*. Market Analysis.
25. OptumCoding. (2025). *2025 Risk Adjustment Coding and HCC Guide*. Reference Materials.
26. Osmosis. (2024). *Active Recall: The Most Effective High-Yield Learning Technique*. Learning Science.
27. PayScale. (2024). *AAPC Certified Risk Adjustment Coder (CRC) Salary*. Compensation Analysis.
28. PubMed. (2024). *Active Recall Strategies Associated with Academic Achievement in Young Adults: A Systematic Review*. Educational Research.
29. PubMed. (2024). *Applying Adult Learning Practices in Medical Education*. Professional Development Research.
30. PubMed Central. (2023). *Feedback in Medical Education: An Evidence-based Guide to Best Practices*. Educational Methodology.
31. RAAPID Inc. (2025). *Risk Adjustment Coding in 2025: Essential Healthcare Guide*. Industry Analysis.
32. ScienceDirect. (2022). *Harnessing the Power of Spaced Repetition Learning and Active Recall for Trainee Education*. Educational Psychology.
33. Sunshine Health. (2024). *Risk Adjustment Coding Guidelines*. Provider Resources.
34. Task Landing. (2024). *How to Land Remote Risk Adjustment Coding Jobs*. Career Development.
35. University of Arizona. (2024). *Active Recall to the Memory Rescue*. Cognitive Science Research.
36. University of California San Diego. (2024). *Spaced Practice*. Learning Methodology Research.
37. Veradigm. (2024). *CMS 2024 Risk Adjustment Model Changes*. Technology Solutions.
38. VLMS Healthcare. (2024). *Essential Medical Coding Scenarios with Answers Explained*. Clinical Education Materials.

39. Wolters Kluwer. (2024). *Risk Adjustment Software | Health Language*. Technology Solutions.
40. Yes-HIM Consulting. (2024). *2024 CMS-HCC Risk Adjustment Model: Review the Changes Here*. Regulatory Analysis.
41. Yes-HIM Consulting. (2024). *Specificity in Documentation & Coding HCCs: Why It's Important*. Documentation Standards.
42. Yoscenario. (2024). *The Power of Scenario-Based eLearning in Medical Education*. Educational Technology.

Made in United States
North Haven, CT
09 August 2025

71505990R00235